Ecological Studies, Vol. 85

Analysis and Synthesis

Edited by

W. D. Billings, Durham, USA
F. Golley, Athens, USA
O. L. Lange, Würzburg, FRG
J. S. Olson, Oak Ridge, USA
H. Remmert, Marburg, FRG

Ecological Studies

Hermann Remmert (Ed.)

The Mosaic-Cycle Concept of Ecosystems

With 91 Figures

Springer-Verlag Berlin Heidelberg New York
London Paris Tokyo Hong Kong Barcelona

Prof. Dr. HERMANN REMMERT
Fachbereich Biologie
der Universität Lahnberge
Karl-von-Frisch-Straße
3550 Marburg/Lahn, Germany

ISBN 3-540-52502-5 Springer-Verlag Berlin Heidelberg New York
ISBN 0-387-52502-5 Springer-Verlag New York Berlin Heidelberg

Library of Congress Cataloging-in-Publication Data. The Mosaic-cycle concept of ecosystems / Hermann Remmert (Ed.) p. cm. – (Ecological studies; v. 85) Based on lectures which were held and discussed at a symposium of the Werner Reimers Stiftung in Bad Homburg. Includes bibliographical references and index. ISBN 3-540-52502-5 (alk. paper). – ISBN 0-387-52502-5 (U.S.: alk. paper) 1. Ecology – Congresses. 2. Cycles – Congresses. I. Remmert, Hermann, 1931– . II. Werner-Reimers Stiftung. III. Series. QH540.M67 1991 574.5 – dc20 90-10269

This work is subject to copyright. All rights are reserved, whether the whole or part of the material is concerned, specifically the rights of translation, reprinting, reuse of illustrations, recitation, broadcasting, reproduction on microfilms or in other ways, and storage in data banks. Duplication of this publication or parts thereof is only permitted under the provisions of the German Copyright Law of September 9, 1965, in its current version and a copyright fee must always be paid. Violations fall under the prosecution act of the German Copyright Law.

© Springer-Verlag Berlin Heidelberg 1991
Printed in Germany

The use of registered names, trademarks, etc. in this publication does not imply, even in the absence of a specific statement, that such names are exempt from the relevant protective laws and regulations and therefore free for general use.

Typesetting: International Typesetters Inc., Makati, Philippines
31/3145(3011)-543210 – Printed on acid-free paper

Preface

The first international congress for ecology took place in 1974 in The Hague, its central theme being "Unifying Concepts in Ecology". In the forefront of discussion at that time were questions of constancy, stability and resilience. Such questions have gone slightly out of fashion and the exceptionally precise and well thought-out concepts of that era are seldom applied nowadays. The present book introduces another unifying concept, the concept of the ecological cycle, or, more precisely, the mosaic-cycle concept of ecology.

The following chapters have their origin in lectures which were held and discussed at a symposium of the Werner Reimers Stiftung in Bad Homburg. The purpose of the symposium was the preparation of this book. Our warmest thanks go to the Reimers Stiftung for their assistance and hospitality. We should also like to express our gratitude to all participants, to those who contributed to the discussion, and above all to those colleagues whose lectures provided, from a variety of aspects, a critical approach to the mosaic-cycle concept.

Marburg, Winter 1990/91   HERMANN REMMERT

Contents

Participants

Contributors to this volume are indicated with an asterisk.

F. BAIRLEIN
Zoologisches Institut der Universität
Weyertal 119
5000 Köln 41, Germany

H. H. BERRY*
Namib-Naukluft Park
Directorate of Nature Conservation
P. O. Box 1592
Swakopmund 9000
Namibia, Africa

E. BEZZEL
Institut für Vogelkunde
Gsteigstraße
8100 Garmisch-Partenkirchen
Germany

H. ELLENBERG
Institut für Weltforstwirtschaft
und Ökologie
Leuschnerstraße 91
2050 Hamburg 80, Germany

J. HAFFNER*
Tommesweg 60
4300 Essen 1, Germany

F. JELTSCH
Fachbereich Physik
Universität Marburg
3550 Marburg, Germany

H. KORN*
Programa Regional de Vida
Silvestre
para Mesomerica y el Caribe
Escuela de Ciencia Ambinentales
Universidad Nacional
Apartado 1350
Heredia, Costa Rica

D. MUELLER-DOMBOIS*
Dept. of Botany
University of Hawaii
Honolulu/Hawaii, USA

V. NICOLAI
Fachbereich Biologie
Lahnberge
Postfach 1929
3550 Marburg, Germany

W. D'OLEIRE
Nationalparkamt
Doktorberg 6
8240 Berchtesgaden, Germany

H. REMMERT*
Fachbereich Biologie
Universität Lahnberge
Postfach 1929
3550 Marburg, Germany

J. REICHHOLF
Zoologische Staatssammlungen
Münchhausenstraße
8000 München, Germany

K. REISE*
Biologische Anstalt Helgoland
Helgoland
2282 List/Sylt, Germany

M. SCHAEFER
Institut für Zoologie II
der Universität
Berliner Str. 28
3400 Göttingen, Germany

W. SCHERZINGER
Guntherstraße 8
8351 St. Oswald, Germany

W. SCHMIDT
Botanisches Institut
der Universität
3400 Göttingen, Germany

W. R. SIEGFRIED*
Percy FitzPatrick Institute
University of Cape Town
Rondebosch 7700
South Africa

U. SOMMER
Max-Planck-Institut für Limnologie

Postfach 165
2320 Plön, Germany

G. VAUK
Norddeutsche Naturschutzakademie
Hof Möhr
3043 Schneverdingen, Germany

M. VOGEL
Akademie für Naturschutz
und Landschaftspflege
Postfach 1261
8229 Laufen/Salzach, Germany

CH. WISSEL*
Fachbereiche Biologie und Physik
Philipps-Universität Marburg
3550 Marburg, Germany

H. ZIERL
Nationalparkamt
Doktorberg 6
8240 Berchtesgaden, Germany

The Mosaic-Cycle Concept of Ecosystems — An Overview

H. REMMERT

1 Introduction

The mosaic-cycle concept was developed in 1938 by Aubreville, using as an example the pristine forests of what was then French West Africa. In the succeeding years the concept was seldom referred to, with a few rare exceptions such as Richards (1981), until it was revived and expanded by Remmert (1985, 1987, 1988a,b). It seems likely that the mosaic-cycle concept is valid for most, if not all ecosystems; the resulting conclusions and predictions are therefore of the greatest importance for ecosystem research. This can best be illustrated in the case of temperate forest ecosystems.

2 The Concept in Forests

The few remaining pristine forest systems in Europe exhibit a cyclic alternation between an optimal phase consisting of trees of roughly the same height and age (usually of a single species) which then deteriorates into a phase of decay. This is succeeded by a phase of rejuvenation which in time becomes an optimal phase. Stages with other tree species (some of them pioneer species) are often interpolated between the phases of decay and rejuvenation. Leibundgut (1982), for example, wrote that in the remaining primary forests of central, eastern and northern Europe a genuine "climax forest" only develops on a portion of the total area and that a continuous process of change within the forest complex results in the presence not only of various stages of development within the climax forest community but also of a variety of stages in the forest succession. Further, in the opinion of Leibundgut, a restriction of the definition primary or pristine forest to the climatically determined climax forest would mean that parts of the forest would have to be alternatingly designated as primary forest and non-primary forest.

A central European primary forest is thus a mosaic of constituent "stones" or areas in which the same cyclic succession of growth and decay is going on; but the cycles are "out of step" or desynchronized. This explains the origin of the term mosaic-cycle concept. Further examples can be found in Leibundgut (1982), Ellenberg (1978), Mayer (1984), and Mayer (1987). All of these authors report

Fachbereich Biologie, Universität Lahnberge, Postfach 1929, 3550 Marburg, Germany

H. Remmert (Ed.)
Ecological Studies Vol. 85
© Springer-Verlag Berlin Heidelberg 1991

the same picture, in which the optimal phase closely resembles the typical European managed forest. The situation can be summarized as follows:

1. The optimal phase consists of trees of more or less the same age; undergrowth of the same tree species is of comparatively little importance. Ebenmann and Persson (1988) from purely theoretical considerations conclude that organisms with continuing, life-long growth tend to live in even-aged populations. This seems to be important for our considerations, as they include not only trees, but corals, fish, and bivalves in our considerations.

2. Accordingly, when the optimal phase deteriorates into a phase of decay it collapses almost simultaneously over the greater part of that particular area. Young plants can now shoot up and the forest reestablishes itself (Fig. 1).

3. The young plants are often not the original tree species, so that the collapse of

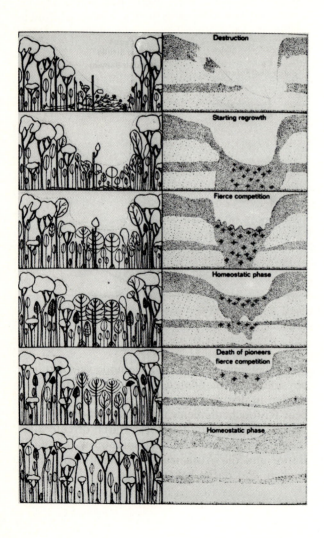

a primary forest, consisting of trees of uniform age, is succeeded by a different tree society which, when it in turn collapses, is replaced by the original primary forest. Figure 2 shows a familiar picture from a different aspect. According to this interpretation, what we encounter in the primary forest is not constancy but rather a more or less regularly occurring cycle of events that is run through asynchronously in the different areas constituting the mosaic-like structure of the ecosystem.

In parts of the Canadian taiga it can be assumed that almost pure spruce forests alternate regularly with almost pure pine forests, each consisting of trees of practically the same age group.

All of these ideas are based on the analysis of different, but neighbouring, forest types. In the deciduous forests of North America, Forcier (1975) used other

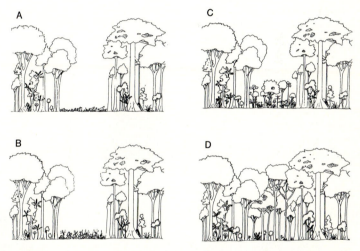

Fig. 2A. Opening in canopy caused by natural tree-fall; diseased tree loses leaves from crown; man-made clearing. **B** After the gap is formed *1* light and temperature increase while humidity decreases; *2* juveniles of pioneer species become established; *3* light-demanding juveniles of forest plants flourish; *4* shade-tolerant understory plants may die. **C** The gap is gradually closed by *1* crowns of trees surrounding gap growing into opening; *2* upward growth of juveniles of secondary species and forest species; *3* climbers and vines growing into the opening. **D** When the gap is closed again *1* low light and temperatures and high humidity are restored in the understory; *2* no more light-demanding seedlings of pioneer plants become established, but adults may persist; *3* seedlings of forest species will become established; *4* understory plants adapted to low light and temperatures thrive again; *5* trees whose crowns form the canopy will not be as tall or as large in girth as trees surrounding the gap (Wong and Ventocilla 1947)

◄───

Fig. 1. Recovery cycle of an African tropical rain forest after the collapse of some of its giant trees. The understory shade plants die off, and a highly competitive situation develops among pioneer tree species, whose canopy then provides the first covering for the gap. Only after this stage do the genuine giant rain forest species reestablish themselves (After Hallé et al. 1978)

methods to support the interpretation that the mosaic-like occurrence of different tree species is the expression of a cyclic process (Fig. 7). Based on his studies of the reproductive strategies of the different tree species he postulated a cycle for the primary forests (Fig. 3) of the temperate zone that has found widespread acceptance in the USA. The optimal phase of *Fagus grandifolia* is followed – sometimes via transitional phases—by the phase of ageing and decay; this in turn is succeeded by birch (*Betula alageniensis*) and then by a mixed forest with sugar maple (*Acer sacharum*), which is finally replaced by *Fagus grandifolia*, thus completing the cycle.

In Germany we have employed a third method, involving physiological measurements, to detect possible cycles in the most commonly occurring forest, the beech forest (*Fagus sylvatica*). We measured the warming of the bark due to direct sunshine, the insulating effect of the bark and the warming of the cambium. The common beech is noted for its inability to tolerate intense irradiation and is susceptible to sunburn in summer (Fig. 8; Nicolai 1986a,b). The bark splits and peels off, which in the end leads to death of the tree (Fig. 4). As a result, the sunrays

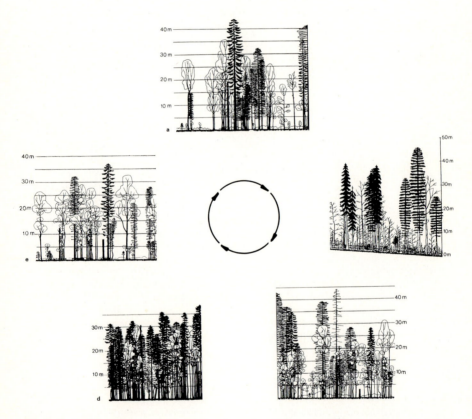

Fig. 3. Cycle of development of a European primary forest (Remmert 1989; using a figure taken from Ellenberg 1978, individual pictures rearranged)

can now reach the trunk of the next tree within the forest and it then suffers a similar fate. In a mature beech forest, windbreak causes the forest to recede further and further, its place is taken mainly by perennials, and then by birch, seeds of which are present in large numbers in the soil of nearly every central European forest. The white bark of the birch trees reflects incoming sunlight almost completely so that overheating of their trunks is negligible. The birch trees are followed by tree species whose bark, at the adult stage, is covered with cracks, such as elm (*Ulmus*), ash (*Fraxinus*), Norway maple and sycamore (*Acer*) and wild cherry (*Prunus*). Cracked bark insulates the phloem much better than the smooth bark of the beech. The trunk of the oak (*Quercus*) has very thick bark, which provides it with additional insulation. Oak and the other above-mentioned species can, unlike the beech, tolerate exposure of their trunks. Very often there is a shortcut in the cycle, and beech follows beech.

A careful scrutinizing of the literature (especially Mayer 1984; Mayer 1987; Ellenberg 1978) reveals numerous cases that substantiate the cycle idea (Fig. 6). Falinski (1988), for example, described many such processes from the primary forest region of Bialowieza in Poland. Figure 5 shows such situations, in which a deteriorating alder forest is in the process of being replaced by young spruce.

Within this cycle of beech, birch and mixed forest, consisting of cherry, ash and maple, there is also a cycle in the species diversity of flora and fauna. Whereas the mature beech forest stage is very poor indeed with regard to species, the phase of decay is accompanied by a dramatic increase in their numbers, a maximum is achieved during the mixed forest phase, after which the number drops again. Thus, the longest phase of the cycle shows the lowest diversity. The plants and fauna of the forest floor, birds, and small and large mammals are also included in the cycle of species diversity. Similar cycles apparently exist in other systems as well.

Many cycles of this kind have by now been described for tropical rain forests, although because of the large numbers of different tree species the mosaic "stones" are smaller and the cycles therefore not so easy to recognize as in the temperate primary forests of central Europe or North America. The numerous studies on tree-fall gaps (Table 1) that have appeared in recent years describe similar situations. When a giant of the rain forest crashes to the ground, it creates a gap in the forest, the shade plants are killed by the sudden influx of light, and the seeds, seedlings and young plants of light-requiring trees already present on the forest floor can now germinate and shoot up at a great pace. These are genuine pioneer plants that will only exist for the relatively short period of little more than

Table 1. The size of the mosaic stones in different forests

Biome	Gap size
Tropical rain forest	71–615 m^2
Beech forest, Europe	0.1–2 ha
Mopani forest, Botswana	0.1–5 km^2

Young phase

Optimal phase

Aging phase

Decay and rejuvenating phase

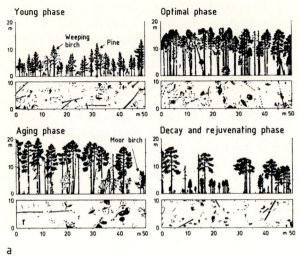

a

ech-primary forest Dobra
ing phase

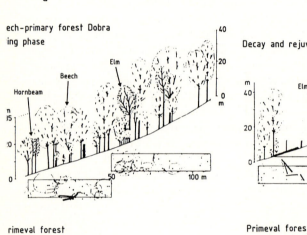

Decay and rejuvenating phase

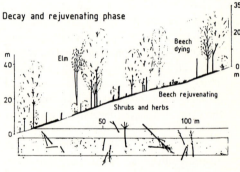

rimeval forest
ustria: Optimal phase

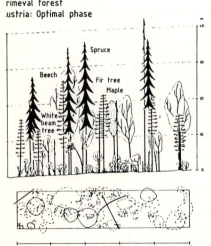

Primeval forest
Austria: Aging phase

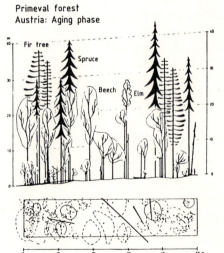

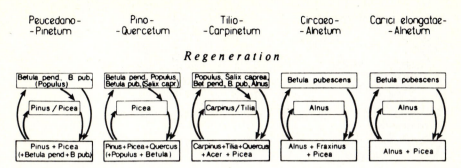

Fig. 5. The Bialowieza primary forest in Poland. No tree is replaced by the same species (Falinski 1986)

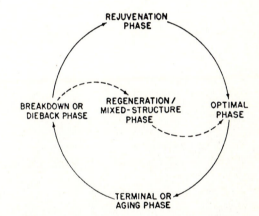

Fig. 6. The forest life-cycle as suggested for European virgin forests. A separate stand reestablishment or building phase is sometimes recognized between the rejuvenation and optimal phase. The differences between rejuvenation and regeneration phases may reflect variations in spatial scale and rates of stand-level breakdown (After Mayer and Neumann from Mueller-Dombois 1987)

Fig. 4. Examples of the forest cycle. **a** Development phase of a primary pine forest in Sweden. Birches are not present in the optimal phase although they play an important role in the juvenile phase (Ellenberg 1978). **b** *Left* Terminal phase of a beech-spruce-pine primary forest on the Balkan peninsula. *Right* Phase of decay and rejuvenation of the same forest. In this phase the number of species is distinctly higher than in the optimal phase (Ellenberg 1978). **c** Optimal phase and Terminal phase of a primary beech forest in lower Austria. In the latter phase an important role is played by perennials and by natural rejuvenation of tree species (After Mayer 1987)

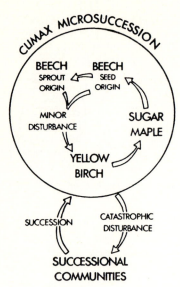

Fig. 7. Succession cycle in North American broad-leaved deciduous regions (After Forcier 1975)

Fig. 8. During the growth season, beech is sensitive to insolation and suffers sunburn. Oak responds by crown death and by putting out new shading shoots in the trunk region

100 years before the giant trees of the rain forest gradually take over again. In the end effect, the tropical rain forest presents the same picture as the forests of the temperate zone, but on a smaller scale and with greater diversity in the tree-fall gaps.

In their dynamics, the species-rich primary forests of eastern Europe, such as the Bialowieza forest, are reminiscent of the tropical rain forests (Fig. 9). The trees die off individually, rather than more or less simultaneously as extensive stands, and the tree-fall gaps are then filled by dense regrowth. Unlike tropical forests, however, these eastern European forests are relatively thin and allow the growth of a good understory. The highest biotic diversity of these multi-species forests is shown by tree-fall gaps. Trees which are uprooted in their fall cause a much higher diversity than trees which rot at their base, thus falling without disturbing the soil (Platt and Strong 1989; Falinski 1988).

A special problem is presented by forests consisting of a single tree species, such as the birch (*Betula*) forests of northern Europe, the beech forests (*Nothofagus*) of New Zealand and South America, European spruce (*Picea*) forests of the upper regions of mountain ranges, and the Mopani forests (*Colophospermum mopane*) of Africa (Fig. 10). In northern Botswana the Mopani forests extend for hundreds of kilometers. Extensive areas are obviously in the optimal phase, equally large areas are occupied by dead trees standing in the grass, and there are also vast open grasslands in which careful search often reveals

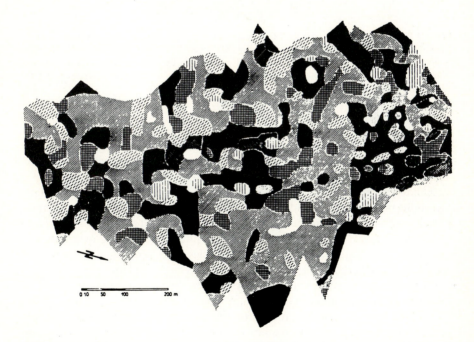

Fig. 9. The spatial distribution of stand-development phases in an undisturbed, 1-km-wide, central core area of a Yugoslavian virgin forest (After Mayer from Remmert 1989)

Fig. 10. The practically one-species Mopane forests of northern Botswana consist of mature, even-aged stands (**above**), which then die off, but may nevertheless remain upright for long periods of time (**middle, left**); when they fall, grass steppe forms (**below**), on which later and at much the same time Mopane trees once again begin to grow (**right**) (Remmert 1989). The biotic diversity is lowest in the mature stand

the stumps of dead Mopane trees. In addition, there are similarly large areas of Mopani saplings, roughly 1m in height, reminiscent of young beech plantations. The hypothesis of Petrides (1974) that the African steppe alternates between savannah and grassland is obviously valid for the Mopani forests. In all of these forests the phase of ageing and decay leads to the formation of areas of grasslands, a phenomenon that in Europe is often wrongly interpreted as indicating the onset of "forest dieback". In fact, this is a normal phase in the cycle of events in a monospecies forest, where the dying tree species is not replaced by another tree species but by a herbaceous plant (Fig. 10). It is even normal for half-grown trees to die suddenly on poor ground, simply because the nutrients are exhausted, or because of the accumulation of too many disease-carrying organisms.

Where cycles of this nature are distributed in a mosaic-like fashion throughout the entire ecosystem, the question of their significance for the system as a whole has to be considered. We also have to look for the driving forces, the causes and the consequences of this phenomenon. When all such aspects have been discussed, we can go on to consider whether the mosaic-cycle concept is also valid for systems such as the tundra or steppes, and whether it also applies to entirely different systems such as the open-water system with its plankton, or the ocean floor with its animal communities. In the following sections we shall take a closer look at these questions.

Presently, it is uncertain whether herbaceous plants in the forest undergrowth also show the mosaic-cycle phenomenon. For the heath (*Calluna vulgaris*) the following concept is valid: old heath dies and is replaced by lichen and grasses and then returns again. Similar conditions are described in the text and shown in particular in the figures of Dierschke (1989) for such communities on the floor of a beech forest. For simplicity, however, I have restricted myself in the following to the description of large, clearly recognizable cycles. Here, too, it must be remembered that part of a cycle can be omitted, and that some parts can be repeated one or more times, which further complicates the picture.

The fact that, contrary to general opinion, primary forests consisting of one or only a few species are made up entirely of individuals of roughly the same age, is often attributed to forest fires. Although this is true in some cases it is by no means the only possible explanation, as is clearly illustrated by the European beech forests.

It is thought that the most diverse plant community in the world, the South African Fynbos, may be a relict of an earlier cycle whose forest phase was cut down by the first Europeans in the area. This theory finds support in place names like Houtebay (timber bay), and old prints of the Cape on which large trees can be distinguished, in addition to the fact that, despite its enormous number of species, the plant life here is under serious threat due to the immigration of plant species of very diverse origin. As a low-species tree phase, the most likely suggestion seems to be the very slowly growing, and today extremely rare "yellow woods" (*Podocarpus*), whose wood was much valued by the earliest European seafarers for repairing their ships.

It might be more correct to speak of a helical process in these long-term cycles since there need not necessarily be a complete return to the starting point. In view of the length of the cycles this would allow for an evolutionary process of some kind.

3 The Significance of the Mosaic-Cycle Concept for an Understanding of Ecosystem Processes

If the mosaic-cycle concept is valid, then there is no such thing as a uniform habitat since after only a short period of time all habitats will assume a mosaic-like structure. The old question as to uniformity or diversity of habitats thus answers itself. The same applies to questions concerning regulatory processes in populations and ecosystems (Fig. 11). The situation can be compared to the biochemical events in an animal organism, in which all particularly important functions are cyclic processes (the Krebs cycle, for example). A less important role is played by linear processes. The same seems to apply in ecology. According to

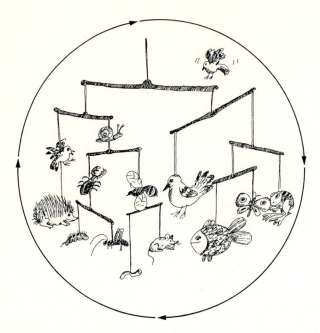

Fig. 11. The generally accepted ecological scheme of cybernetic interaction needs to be expanded to include larger (regulatory) cycles (Remmert 1989)

the mosaic-cycle concept, equilibrium can be maintained for the most part without intraspecific self-regulatory feedback processes. In every case, the system progresses in one direction, towards a catastrophic event such as the fall of a giant tropical rain forest tree, with the subsequent death of the shade plants growing below it. Every imaginable catastrophe that can take place in an ecosystem, as well as the regeneration after such catastrophes, is already contained within (or provided for by) the system's program. Ecological equilibrium would then be replaced by desynchronized cycles.

The massive multiplication of pest insects, of nematodes, fungi, viruses or other plagues such as have been frequently observed in near-natural systems (in the terminal phase) could thus be regarded as part of a system which automatically includes the compensation of these events. An example of this kind is the well-known case of the spruce budworm (*Choristoneura fumiferana*; Insecta; Lepidoptera) which is responsible for the large-scale alternation of spruce and pine in the Canadian taiga. It is possible that the multi-stable systems that are attracting so much attention at present might better be interpreted as parts of cycles.

It seems that for the maintenance of a long-term average niveau in an ecosystem, "feedback" processes are of little significance, cyclic processes are less susceptible to interference and play a much more important role.

Such considerations are particularly relevant in theoretical discussions concerning species number and diversity. The end phase of a natural vegetation, its climax, is thus apparently a mosaic of different plant communities, each independently running through its own cycle. Certain phases of the cycle — for example the optimal phase in a beech forest — are poor in species (the system is dominated by a single plant species and the faunal diversity is also low), whereas at another phase of the same system there may be a large number of animal and plant species. The same holds true for the tropical rain forest: the greatest number of species is found in clearings caused by the fall of a giant tree. Thus within one and the same system high and low diversity alternate with one another (Fig. 14).

4 The Driving Forces of the Cycle

The forces driving a cycle of events in the ecosystem are in fact none other than the potential longevity of its constituents or, in other words, the average age attainable by beech (*Fagus sylvatica*), birch (*Betula pendula*) and maple species (*Acer*). However, these forces can be considerably modulated by other effects. Windbreak (storms), disease, or the large-scale multiplication of injurious insects may greatly accelerate certain stages of the cycle and thus lead to shorter cycles. On the other hand, some stages are not necessarily succeeded by a stage with different tree species as would normally be expected; in a beech forest for example, the same stage may be repeated several times, thus slowing down the cycle.

The beaver (*Castor fiber*) is a good example of an animal that initiates cycles of this kind in the forests of the northern temperate zone (Fig. 12). On gently undulating terrain the animal builds dams across streams, large lakes are formed and the trees die off. Due to the high productivity of these shallow lakes, a layer of decaying mud is produced so that the lake silts up within a relatively short

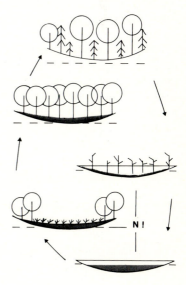

Fig. 12. In the primary forests on the gently undulating country of North America the dam-building activities of beavers lead to the formation of lakes, in which the trees die. When, subsequently, the lake silts up, it gives way to rich grassland which in turn is invaded by trees such as willow, alder and poplar (Remmert 1985), which later give place to "climax" species

period of time. During this time, the fixation of nitrogen from the air is much greater than that taking place in the soil of the surrounding forest (Naiman and Mellilo 1984). If the lake has formed over a foundation of sandstone, it is quite possible that several meters of humus containing very high reserves of nitrogen will have formed by the time the lake has silted up. The region is then rapidly colonized by soft woods and, from the edge of the forest, by other tree species. The humus layer is used up and the original type of forest is reestablished. But this, too, can be destroyed by another beaver lake and thus, in cyclic succession, phases of forest on impoverished soil, beaver lake, beaver meadow, softwood water-meadow on rich soil, forest on rich soil and then forest on impoverished soil follow one after the other.

5 Causes of the Cycle

The underlying cause of the cycles is generally held to be competition for essential nutrients, including light (Tilman's theory of competition: cf. Sommer, this Vol.).

Different plant species do not always have identical mineral requirements and in some cases even differ with respect to their ability to extract minerals from the soil (cf. Killingbeck and Costigan 1988; Tilmann 1989). If a particular tree species has removed essential nutrients from a certain site for centuries, as sometimes happens, the same species cannot immediately thrive again on the same site but must wait until the old trunks have been remineralized (e.g. Uhl 1987; Tilman 1989). On nutrient-rich soil the situation may be different from that on poor soil, although in the long run the principle is the same.

Exactly the same situation exists with respect to competition for light. Beneath trees that give plentiful shade (the common beech in central Europe for example), satisfactory growth only becomes possible when some form of interference results in gaps in the leaf canopy, enabling light to penetrate to the forest floor. The young plants of light-requiring trees like the beech are unable to mature in the shade.

According to Harper (1977), the frequently observed failure of a plant, under natural competitive conditions, to thrive on the same site as the previous generation is a widespread phenomenon. In addition to a selective reduction of resources due to the accumulation of specific plants, he suggests that microorganisms or allelopathic effects may also be responsible. Each of these factors has been observed to play a role in the case of forest trees and many other plants.

6 Consequences Arising from This Situation

The cycles described above lead to a number of obvious consequences, most of which have so far been neglected.

1. Equilibrium will not be encountered within a single mosaic unit, but is rather a unidirectional process. Since the areas chosen for ecosystem studies have to be

as similar as possible, it is in any case obvious that equilibrium will not be observed.

2. The failure of the predominating tree species in a forest to replenish itself need not necessarily be an alarm signal but may be quite natural. We have already mentioned that self-replenishment of the European beech is hardly possible under the dense crown canopy. In principle, this is universally the case. Normally, the umbrella acacias of the African savannah do not seem to be able to rejuvenate on the same site. All of the trees of this species that are visible over any one area are of approximately the same age and only rarely are younger trees seen among them. The young acacias form thickets in other places, and are especially attractive to antelopes.

3. A normal population pyramid for the forest trees will only be found if a population census is carried out over very large areas. As a rule, however, as Fig. 13 shows, only sections of a pyramid are found in the individual mosaic patches: in some places there will be only tree saplings, in other places — such as the mature stage of a beech forest — all trees are 80 to 100 years old, and in still other places only old and dying trees are seen.

These disrupted population pyramids may provide useful indicators for the "key species" representing the driving force of the cycles. Only the relatively short-lived species, such as the song birds or smaller mammals in a forest, have normal population pyramids. This may also apply to the relatively short-lived perennials. In any case population pyramids of other habitats will have to be re-analyzed. Mertens's (1988) finding of totally disrupted population pyramids for green frogs (*Rana esculenta* groups) in a German pond system would seem to support this idea. Pelagic fish in lakes and in the sea live in even-aged swarms. It seems possible that these size-structured swarms may be described as a mosaic cycle of the whole population in a given system.

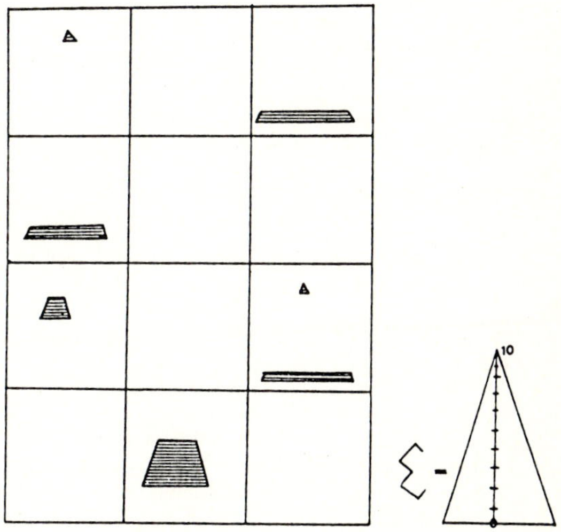

Fig. 13. For the trees in a uniform mosaic unit a normal population pyramid cannot be constructed. In each mosaic unit there is only a fragment, or perhaps no specimen at all, of the normal population pyramid present

4. In all likelihood, failure to recognize and respect these cycles is in part responsible for the problems encountered in forests today (Fig. 14). Planting the same tree species generation upon generation, and on soil that is dense with roots of the same species will in any case aggravate the situation: the problems are bound to increase with each succeeding generation since it is always the same nutrients that are removed from the soil. Furthermore, the previous generation of trees were felled at the optimal phase, when the roots were at a maximum.

5. The soil of such a system must also be subject to cyclic changes. The ground is so thoroughly permeated by the roots of the dominant tree species that root competition alone would make it highly unlikely that a different species, or young plants of the same species, would be able to survive. In fact, only often the phase of ageing and decay, when the root density lessens, can colonization with new plants be expected. Another factor to be considered is that the leaves of different trees decay at very different rates. The litter in our example of the European beech forest consists, for several hundreds of years, of leaves that are very resistant to decay. This is succeeded by a short phase with decaying perennials, followed by birch leaves and finally by the usually fast-decaying leaves of a mixed forest of elm, ash and maple. It seems very likely, therefore, that the different stages of the cycle will produce quite different kinds of soil and that soil fauna and flora will also differ from stage to stage.

A changeover from one tree species to another — as seen on the Kaibab plateau — is also accompanied by strong fluctuations in the acidity of the soil.

6. In addition, it is a well-known fact that a living forest consumes vast quantities of water, with the result that the groundwater level drops considerably and only

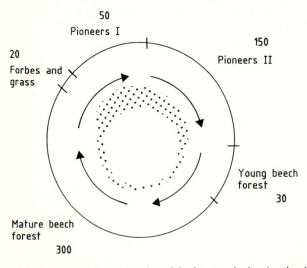

Fig. 14. Schematic representation of the forest cycle showing the changing diversity of the system. Even if there is a short cut in the cycle and the beech forest is replaced by a new beech forest after the ageing stage, the cycle in our sense remains the same. The *figures* indicate the normal time of each phase in this cycle

rises again when the forest disappears. This has been observed time and again following clear-felling. Such fluctuations are of course accompanied by marked changes in the fauna. Herrchen (1989) showed that in windbreaks in the primary forests of the Bavarian Forest National Park not only was there a high ground-water level, but the rising water even formed shallow ponds. This brought about a change in the fauna: the bank voles and yellow-necked mice that were the most numerous small mammals in the dry forest were suddenly joined by water shrews on the windbreaks.

This brings up yet another question, concerning the way in which the newly formed small units in the mosaic are colonized. How do plants and animals find these favourable places?

7. The cycle concept holds not only for the end phase in the development of the plants, but also applies to the pioneer succession from the pioneers onwards. Most *Calluna* heaths in central Europe exhibit such stages on the way to the end stage: a clearly recognizable phase of growth, an optimal phase, and a stage of death and disintegration in which lichens and Poaceae partially cover the ground of the bare patches.

8. At this point a question of terminology arises, although it is concerned with more than just terminology. An ecosystem is in fact defined by human criteria and not by genuinely scientific considerations. While the situation in the forest is relatively simple (it is difficult to regard the optimal phase of a beech forest as merely a part of a cycle instead of an ecosystem of its own), it is much more complicated in the case of herbaceous plants and perennials. In fact, the system is the same — whether in the steppe or the treeless tundra: an old perennial dies, other plants appear in its place and not until some time has elapsed can a plant of the original species again grow on the old site. This happens in an area of a few square meters, and so far no one has been tempted to describe phases of this kind as independent ecosystems. The same applies to certain phases in the development of a plankton community (cf. Sommer, this Vol.). Clearly, an ecosystem is a system created solely on the basis of subjective phenomena and is not an entity defined and delineated by scientific criteria.

7 Colonization of the Mosaic Units

At their imago stage, insects, of which many species are specific for a particular phase of the cycle (characteristic for the tree-fall gaps in a tropical rain forest), appear to be continuously on the look out for suitable places to mate and deposit their eggs. Seeds or rhizomes of some plants are known to lie inactive in the soil for as long as 240 years before their growth is stimulated by sudden warmth from a fire caused by lightening. An example is provided by the genus *Terminalia* (Geldenhuis, pers. commun.). The rhizomes of the fern *Pteridium* have also been shown to survive 240 years in the soil of the South African Fynbos, without the appearance of any above-ground parts. Relatively little is known about these examples and much work must still be done.

Seen in this light, prolific seed production throughout the long life of a tree often appears to be pointless, whereas the strategy of the many tropical trees that bear fruit only once, and this towards the end of their life, seems to be the better one.

Our hypothesis also helps in estimating the numbers of large animal species in a closed primary forest. In the present-day managed forests of North America and Eurasia the relicts of large game species are responsible for quite considerable damage. This has led to the assumption that the natural density of large game must be very low. However, if we take as an example the European forests, from which wild horses, aurochs, bison and moose have disappeared, and calculate from the commonly accepted density of between 0.5 and 1 head of large game per km^2, then for a total of six species of large game (wild boar, red deer, moose, bison, aurochs, horse) we obtain a figure of roughly one individual of each species per 10 km^2. Considering the social behaviour of the animals, such a figure is unrealistic and, consequently, a much higher density would have to be expected in a natural primary forest. A mosaic-like forest structure could accommodate much higher densities with a minimum of damage, although this would imply an extremely unequal distribution of the large animals. They would be found in great numbers in and near the meadows and soft-wood areas that are found in every cycle, but they would be practically absent from the optimal phase shortly before its collapse. Studies carried out in the tropical rain forests of Central and South America on a somewhat richer type of soil (in the Andes) lend support to predictions of this kind.

8 General Validity of the Concept

The principle of desynchronized cycles as the structural units in a mosaic-like excosystem apparently applies not only to forests. Petrides (1974) suggested such a system for the African steppe. Similarly, in the Peninsula Valdes National Park (Argentina), there appears to be an alternation of grassy steppe and thornbush savannah and of their corresponding fauna, with guanacos inhabiting the thornbush savannah, nandus the grassy steppe. In Iceland (Remmert 1989), a swan (*Cygnus*) biotope with cotton grass apparently alternates with a grass-meadow biotope of the short-beaked goose. The vegetation of treeless regions such as salt meadows, steppes and tundra usually has a mosaic-like appearance, the probable explanation being a mosaic of non-synchronized cycles. On *Calluna* heaths or salt meadows the presence of desynchronized cycles has been well-documented and analyzed (on a m^2 scale). In the Mongolian steppe the energy and mass turnover is to a large extent determined by small burrowing mammals. The colonies of *Microtus brandti* are so dense that the vegetation is exhausted after a time and the animals move elsewhere. Since the abandoned sites are not only well drained and aerated but also contain large quantities of excrement, they soon support a diversity of aromatic plants; these are gradually replaced by plentiful growth of plants suitable for warm-blooded grazing animals and herbivorous insects. Productivity then gradually sinks to a relatively low level, the mice return

Fig. 15. Kelp species on the Californian coast exhibit the same mosaic cycle with openings and mature stands as do virgin forests (Dayton et al. 1984)

and the cycle begins again. A similar situation is caused by the rodents (*Ctenomys*) in South America and by prairie dogs (*Cynomys*) in the North American prairie. Polish ecologists estimate that roughly 40% of the Mongolian steppe is affected by cycles of this type, although only about 2% of the total steppe is actually inhabited by the mice colonies. The fact that the members of the most important animal species in marine bottom communities are represented by individuals of the same age class has been pointed out by Powell and Cummins (1985), Valiela (1984), and Reise (1981, 1985). The successive phases of Lake Nakuru in Kenya differ strikingly from one another (Vareschi and Jacobs 1985). There is good evidence that the situation is very similar in the "meadows" of the giant kelp, *Macrocystis*, along the Californian coast (Fig. 15; Dayton et al. 1984). My assumption is that each of the cases described above involves a general principle for the regulation of ecosystems, by which — because catastrophes and their compensation are built into the system — damage and disturbance to the ecosystem can be much more easily corrected than would be possible by a cybernetic "feedback" system of different organisms. More will be said about this in the following contributions.

References

Andreev A (1988) The ten year cycle of the willow grouse of Lower Kolyma. Oecologia 76:261–267

Archer S, Scifres C, Basshkam AC, Maggio R (1988) Autogene succession in a substropical savanna: conversion of grassland to thorn woodland. Ecol Monogr 58(2):111–127

Aubreville A (1936) La foret coloniale: Les forets de L'Afrique occidentale francaise. Ann Acad Sci Colon Paris 9:1–245

Augspurger CK (1988) Input of wind-dispersed seeds into light-gaps and forest sites in a Neotropical forest. J Trop Ecol 4:239–252

Barton AM, Fetcher N, Redhead S (1989) The relationships between treefall gap size and light flux in a Neotropical rain forest in Costa Rica. J Trop Ecol 5:437–439

Borman FH, Likens GE (1979) Patterns and process in a forested ecosystem. Springer, Berlin Heidelberg New York

Connell JH (1978) Diversity in tropical rain forests and coral reefs. Science 199:1302–1309

Dayton PK, Currie V, Gerrodette T, Keller BD, Rosenthal R, Tresca DV, (1984) Patch dynamics and stability of some california kelp communities. Ecol Monogr 54(3):253–289

Dierschke H (1989) Kleinräumige Vegetationsstruktur und phenonologischer Rhythmus eines Kalkbuchenwaldes. Verh Ges Ökol, Göttingen 1987, B 17

Ebenman B, Persson L (eds) (1988) Size-structured populations. Springer, Berlin Heidelberg New York Tokyo

Ellenberg H (1978) Vegetation Mitteleuropas mit den Alpen, Ulmer, Stuttgart, 981 pp

Falinski JB (1986) Vegetation dynamics in temperate lowland primeval forests. Junk, Dordrecht, 537 pp

Falinski JB (1988) Succession, regeneration and fluctuation in the Bialowieza Forest (NE Poland). Vegetatio 77:115–128

Forcier LK (1975) Reproductive strategies in the co-occurrence of climax tree species. Science 189:808–810

Gerrish G, Mueller-Dombois D, Bridges KW (1988) Nutrient limitation and *Metrosideros* forest dieback in Hawaii. Ecology 69(3):723–727

Glitzenstein JS, Harcombe PA, Streng DR (1986) Disturbance, succession, and maintenace of species diversity in an east Texas forest. Ecol Monogr 56(3):243–258

Grubb PJ (1977) The maintenance of species richness in plant communities: the importance of the regeneration niche. Biol Rev 52:107–145

Hallè F, Oldeman RAA, Tomlinson PB (1978) Tropical trees and forests. An architectural analysis. Springer, Berlin Heidelberg New York, 441 pp

Harper J (1977) Population biology of plants. Academic Press, London, 892 pp

Harrison S (1986) Treefall gaps versus forest understory as environments for defoliating moth on tropical forest shrubs. Oecologia 72:65–68

Herrchen S (1989) Ökologische Untersuchungen an Kleinsäugern auf Windwurfflächen des frühen Sukzessionsstadiums im Nationalpark Bayerischer Wald. Diplomarbeit, Marburg

Jacobs M (1988) The tropical rain forest. Springer, Berlin Heidelberg New York Tokyo, 295 pp

Jones EW (1945) The structure and reproduction of the virgin forest of the north temperate zone. New Phythol 44:130–148

Killingbeck KT, Costigan SA (1988) Element resorption in a guild of understory shrub species: niche differentiation and resorption thresholds. OIKOS 53:366–374

Leibundgut H (1982) Europäische Urwälder der Bergstufe. Haupt, Bern, 306 pp

Leigh EG, Rand AS, Windsor DM (eds) (1982) The ecology of a tropical forest. Smithsonian Inst Press, Washington

Mayer H (1984) Wälder Europas. Fischer, Stuttgart, 691 pp

Mayer H (1987) Urwaldreste, Naturwaldreservate und schützenswerte Naturwälder in Österreich. Wien, Univ Boden Kultur, 971 pp

Merkle J (1954) An analysis of the spruce-fir community on the kaibab plateau, Arizona. Ecology 35(3):316–322

Mertens D (1988) Populationsökologische Untersuchungen an den Wasserfröschen des neuen Botanischen Gartens der Marburger Philipps-Universität. Diplomarbeit, Marburg

Mueller-Dombois D (1983a) Population death in Hawaiian plant communities: a causal theory and its successional significance. Tuexenia 3:117–130

Mueller-Dombois D (ed) (1983b) Forest dieback in Pacific forests. Science 37(4):313–496

Mueller-Dombois D (ed) (1983c) Canopy dieback and dynamic processes in Pacific forests. Pac Sci 37:313–496

Mueller-Dombois D (1984) Zum Baumgruppensterben in pazifischen Inselwäldern. Phytocoenologia 12(1):1–8

Mueller-Dombois D (1985) Ohi'a dieback in Hawaii: 1984 synthesis and evaluation. Pac Sci 39(2):150–170

Mueller-Dombois D (1987) Natural dieback in forests. BioScience 37(8):575–583

Mueller-Dombois D (1988a) Towards a unifying theory for stand-level dieback. GeoJournal 17(2):249–251

Mueller-Dombois D (1988b) Forest decline and dieback — a global ecological problem. TREE 3:310–312

Naiman RJ, Mellilo JM (1984) Nitrogene budget of a subarctic stream altered by beaver (Castor canadensis). Oecologia 62:150–155

Nicolai V (1986a) Selbst Bäume schützen sich vor Sonnenbrand. Forschung Mitt DFG 1:4–6

Nicolai V (1986b) The bark of trees: thermal properties, microclimate and fauna. Oecologia (Berlin) 69:148–160

Petrides GA (1974) The overgrazing cycle as a characteristic of tropical savannas and grasslands in Africa. Proc 1st Int Congr Ecol 86–91, Wageningen

Platt WJ (ed) (1989) Gaps in forest ecology. Ecology 70:3

Platt WJ, Strong DR (eds) (1989) Special feature: gaps in forest ecology. Ecology 70(3):535–570

Popma J, Bongers F, Martinez-Ramos, Veneklaas E (1988) Pioneer species distribution in treefall gaps in Neotropical rain forest; a gap definition and its consequences. J Trop Ecol 4:77–88

Powell EN, Cummins H (1985) Are molluscan maximum life spans determined by long-term cycles in benthic communities? Oecologia 67:177–182

Reise K (1981) Ökologische Experimente zur Dynamik und Vielfalt der Bodenfauna in den Nordseewatten. Verh d DZG 1981, Fischer, Stuttgart, pp1–15

Reise K (1985) Tidal flat ecology. Ecol Stud vol 54, Springer, Berlin Heidelberg New York Tokyo

Remmert H (1985) Was geschieht im Klimax-Stadium? Naturwissenschaften 72:505–512

Remmert H (1987) Sukzessionen im Klimax-System. Verh Ges Ökol Gießen, B XVI:27–34

Remmert H (1988a) Wie verjüngt sich ein Urwald? alma mater philippina, Marburger Universitätsbund, 4–7:7

Remmert H (1988b) Gleichgewicht durch Katastrophen. Aus Forschung und Medizin 1:7–17

Remmert H (1989) Ökologie. Springer, Berling Heidelberg New York Tokyo

Richards PW (1981) The tropical rain forest. Cambridge Univ Press, 450 pp

Rowell CHF, Rowell-Rahier M, Braker HJ, Cooper-Driver G, Gomez LD (1984) The palatability of ferns and the ecology of two tropical grasshoppers. Biotropica 15:207–216

Schaller GB, Jinchu H, Wenshi P, Jing Z (1985) The giant pandas of wolong. Univ Chicago Press, 298 pp

Scherzinger W (1986) Die Vogelwelt der Urwaldgebiete im Inneren Bayerischen Wald. Schriftenreihe Bayerischen Staatsministeriums Ernährung, Landwirtschaft und Forsten 12:188

Schupp EW (1988) Seed and early seedling predation in the forest understory and in treefall gaps. OIKOS 51:51–78

Shugart HH (1987) Tree death: cause and consequence. BioScience 37:540–609

Sprugel DG (1976) Dynamic structure of wave-regenerated Abies balsamea forests in the northeastern United States. J Ecol 64:889–911

Steven de D (1988) Light gaps and long-term seedling performance of a Neotropical canopy tree. J Trop Ecol 4:407–411

Swaine MD, Hall JB (1988) The mosaic theory of forest regeneration and the determination of forest composition in Ghana. J Trop Ecol 4:253–269

Tilman D (1989) Competition, nutrient reduction and the competitive neighbourhood of a bunchgrass. Funct Ecol 3:215–219

Uhl C (1987) Factors controlling succession following slash-and-burn agriculture in Amazonia. J Ecol 75:377–407

Valiela I (1984) Marine ecological processes. Springer, Berlin Heidelberg New York, 546 pp

Vareschi E, Jacobs J (1985) The ecology of Lake Nakuru. Oecologia 65:412–424

White TCR (1986) Weather, Eucalyptus, dieback in New England and a general hypothesis for the canopy dieback. Pac Sci 40:58–78

Wong M, Ventocilla J (1947) A day on Barro Colorado Island. Smithsonian Inst, Balboa, 93 pp

Woodroffe CD (1988) Relict mangrove stand on last interglacial terrace, Christmas Island, Indian Ocean. J Trop Ecol 4:1–17

A Model for the Mosaic-Cycle Concept

Ch. Wissel

1 Introduction

In many empirical investigations of ecosystems only a small part of the system is considered for a short period of time. In the interpretation of data obtained in this way equilibrium considerations are often stressed and no spatial heterogeneity is taken into account. The idea of the climax state fits into this approach. It describes the final state of a succession which should be determined by the local climatic and physical conditions. It is mainly believed to be an equilibrium state and its spatial pattern is seldom discussed. But there are investigations in which a climax is seen which is not in a constant equilibrium state, but rather a cyclic one (Mayer and Neumann 1981; Burton and Mueller-Dombois 1984; Sprugel 1976; Putman and Wratten 1984; Yeaton 1978). It is also well known that ecosystems may show a typical spatial pattern.

The situation in theoretical ecology corresponds with this usual empirical approach. Mainly equilibria without a spatial pattern are considered (May 1973; Pimm 1982; Hallam and Levin 1986; Wissel 1989; Yodzis 1989). High complex models which try to give a realistic description of whole ecosystems are not discussed here. They contain so many factors with hundreds of parameters that is impossible to appreciate all the model assumptions. These models cannot give an understanding of the system because it is impossible to assess which detail of the model is responsible for a particular result. In addition, they are so developed to describe a very specific situation only. Therefore, they cannot provide general results. Thus, I do not consider these complex models in theoretical ecology.

The aim of theoretical ecology (Wissel 1989) is to provide an understanding of the mechanisms and functional relations in ecology, to find general statements. Therefore, simple models are used which do not contain all the details of the system. Only key factors are included which are essential for the particular question under consideration.

Such models for communities and ecosystems usually start with a description of the dynamics of the populations or guilds of populations present in the system (May 1973; Pimm 1982; Wissel 1989; Yodzis 1989). Mainly conditions for constant equilibrium are investigated, although models of a few populations are known which include the coexistence of populations with fluctuating individual numbers (Hassell 1978; Hallam and Levin 1986; Wissel 1989). Those ecosystem

Fachbereiche Biologie und Physik der Philipps-Universität Marburg, 3550 Marburg, Germany

H. Remmert (Ed.)
Ecological Studies Vol. 85
© Springer-Verlag Berlin Heidelberg 1991

models do not include the description of spatial patterns. In the case of few populations models, which calculate spatial patterns, are also rare (Hassell 1978; Wissel 1989; Murray 1989).

Indeed, for ecosystems the modelling of the dynamics and spatial dependence of all the populations is too complex to provide a simple, understandable model. Thus, in the present chapter a different approach will be introduced. In accordance with the statements above the questions to be answered by the model are stated first, then the decisive factors are modelled.

2 Mosaic-Cycles in the Middle European Beech Forest

The starting point of our model is the mosaic-cycle concept. Its general idea is described by Remmert (this Vol.). The following model will describe the special situation of the Middle European beech forest and will consider the logical consequences of the mosaic-cycle concept in this case. However, it should be emphasized that the simplicity of the model makes it possible to make general inferences which are also valid for other situations.

2.1 The Concept

Here, we will repeat the points of the mosaic-cycle concept for the Middle European beech forest which are essential for our model (Remmert 1985, 1987). Let us subdivide the forest system into different patches, the mosaic stones. (I call them patches for reasons which will become obvious later on.) In every patch the following cycle sequence takes place (see Fig. 1). When an old beech tree has fallen, an opening remains for about 20 years. This is normally colonized by birch trees. They are there for about 50 years. A mixed forest with different species of trees is then established, lasting approximately 150 years, followed by a thicket of young beech trees. Over 30 years they develop into large trees, which may stand

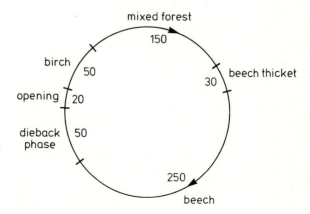

Fig. 1. The cycle of the Middle European beech forest. The *numbers* are the durations of the corresponding phases in years (After Remmert 1987)

there for 300 years. When they fall, the cycle starts again, running independently in every patch (see Fig. 1), i.e. the cycles of the different patches run desynchronously. Thus, at every instant of time the different phases of the cycle are randomly distributed over the patches like a mosaic. It is obvious that this cycle can run only if sufficient seeds of the different tree species are always present for the repeated colonizations.

It is not the purpose of this chapter to discuss the possible causes for this cycle. It is accepted as a concept based on empirical data. The aim of the model is to deduce logical consequences of the mosaic-cycle concept. Here, the influence of one patch on another is of decisive importance.

One possible influence is that an opening may be colonized earlier by birch trees if one of the neighbouring patches is occupied by these trees. If in the neighbouring patches there is no birch tree, but rather mixed wood, the patch under consideration may be colonized earlier by the corresponding tree species. In this case the birch phase of the cycle is skipped.

If a beech tree falls, an opening arises. This may allow strong solar radiation to fall on the trunk of the beech trees in the northern neighbourhood. It is known (Nicolai 1986; Remmert 1987) that this radiation is damaging to beech trees and they will die after a certain time. Thus the opening will proceed northward.

Our model is based on these postulations which are deduced from empirical facts. The aim of the model is to deduce the logical consequences of these postulations. The following questions will be investigated: Are different phases of the cycle really randomly distributed among the patches or will special spatial patterns appear? What are the characteristics of these patterns if they exist? Can the cyclic behaviour also be seen in quantities which describe the whole system, i.e. all patches together? Is there any interrelation between this cyclic behaviour and the spatial pattern? What can be said about the ecological stability of this system? How does it react to different perturbations? What is the influence of the system's size on all of these properties?

2.2 The Model

Let us consider a rectangular forest system subdivided into squares as shown in Fig. 2. The number of squares in the horizontal direction is N_h, in the vertical N_v. Thus, we have $N = N_h \cdot N_v$ squares altogether. In the following $N_v = 42$ and $N_h = 24$ are chosen unless given otherwise. The position of a square is given by the pair of numbers (i,j) with $i = 1,\ldots N_h$ and $j = 1,\ldots N_v$, i.e. (i,j) is the square in the ith line and jth column. The state of the square (i,j) is described by Z(i,j). In the following we measure the time t in units of 10 years. According to Fig. 1, we can identify the phase in every square in the following way:

$1 \leq Z(i,j) \leq 2$:   There is an opening in square (i,j);

$3 \leq Z(i,j) \leq 7$:   There are birch trees in square (i,j) of an age of [Z(i,j)−2] decades;

$8 \leq Z(i,j) \leq 22$: There is a mixed forest in square (i,j) of an age of [Z(i,j)−7] decades;

Fig. 2. Random distribution of the cycle phases (10-years time steps) of Fig. 1 on 1008 squares. See Fig. 3 for explanation of the squares

$23 \leq Z(i,j) \leq 25$: There is a beech thicket in square (i,j) of an age of $[Z(i,j)-22]$ decades;

$26 \leq Z(i,j) \leq 50$: There are adult beech trees in square (i,j) of an age of $[Z(i,j)-22]$ decades;

$51 \leq Z(i,j) \leq 55$: There are old beech trees in square (i,j) of an age of $[Z(i,j)-22]$ decades which may die soon.

The phases of the squares are indicated in Fig. 2. The meaning of the squares are explained in Fig. 3. The darker a square is, the older the beech trees.

After each time step of 10 years the value of $Z(i,j)$ is enhanced by 1, i.e.

$$Z(i,j) \to Z(i,j) + 1 \quad \text{for } Z(i,j) < 55; \tag{1}$$
$$Z(i,j) \to 1 \quad \text{for } Z(i,j) = 55, \tag{2}$$

i.e. a cycle of 550 years is implemented in each square.

This approach differs completely from the usual modelling of the dynamics of ecological systems. Normally, the models are constructed to deduce the local dynamics (Hassell 1978; Hallam and Levin 1986; Wissel 1989; Yodzis 1989). Here, we simply adopt the dynamics from the mosaic-cycle concept. No local regulation has to be modelled. It is not the aim of our model to deduce a reason

Fig. 3. Explanation of the squares. From *left* to *right*: Opening, birch, mixed forest, beech with increasing age from left to right (the darker the square, the older the beech trees)

for this cycle. Our model is similar to the well-known patch models (Christiansen and Fenchel 1977; Wissel 1989), but here the particular position in the system is modelled. Below we will take into account the spatial character of the interactions of the squares which have been introduced above. This approach is similar to cellular automata (Wolfram 1986).

Next, the influences described above have to be included which modify this cyclic behaviour and cause an interaction of the squares. If $Z(i,j) = 1$ for a square and one of its eight neighbour squares contains birch trees, an early colonization will take place. Thus, after one time step the value of $Z(i,j)$ will be

$$Z(i,j) \rightarrow 3 \tag{3}$$

(young birch) instead of 2 (old opening).

If there are no birch trees, but mixed forest in the neighbourhood, a colonization of the opening by the latter appears. Thus, after one time step the value of $Z(i,j)$ will be 8 (young mixed forest)

$$Z(i,j) \rightarrow 8 \tag{4}$$

instead of 2 (old opening).

We call the last part $[51 \leq Z(i,j) \leq 55]$ of the cycle the dieback phase because very old beech trees are endangered by several factors which appear randomly. An old tree may die sooner or later, depending on several factors, e.g. when disturbances like a storm are included. We introduce the probability of dying, P_0, which means that in a square of this phase the value of $Z(i,j)$ is not increased by 1 in one time step in any case, but will pass into $Z(i,j) = 1$, i.e. the beech will die with the probability P_0.

Finally, we model the influence of solar radiation. If an old beech tree falls, the neighbouring beech trees to the north are suddenly exposed to solar radiation and will probably die in about 10 years. The probability P_w of this dying is highest if the solar radiation comes from the southwest. But radiation from the south or southeast can also cause radiation death (Nicolai 1989, pers. commun.). The corresponding probabilities are called P_s and P_e. If north is represented at the top of Fig. 2, we obtain the following transition rules: In a square (i,j) with beech trees the transition

$$Z(i,j) \rightarrow 1 \tag{5}$$

takes place with the probability
P_s if $Z(i, j + 1) = 1$ or 2;
P_w if $Z(i - 1, j + 1) = 1$ or 2; $\qquad\qquad$ (6)
P_e if $Z(i + 1, j + 1) = 1$ or 2.

The transition (1) instead of (5) occurs with the probability $(1-P_s)(1-P_w)(1-P_e)$ if there is no opening in the southern direction, i.e. if the conditions (6) are not fulfilled, the normal transition (1) takes place instead of (5).

At the boundary ($i = 1$ or N_h) or ($j = 1$ or N_v) the patches do not have neighbours on one side. From this side no colonization (3) and (4) can be induced. At the southern boundary the beech trees are exposed to solar radiation. However, as they have grown up in this situation, their trunk should be protected, either by the development of lower foliage or by shrubs in front of them.

It is also possible to mimic an infinite system without boundaries by choosing periodic boundary conditions. That means we take the lower boundary as neighbour for the upper one and vice versa and the right one as neighbour for the left one and vice versa. For instance the patch $(0,j)$ is taken as equivalent to $(N_h + 1,j)$, and $(i,0)$ equivalent to $(i,N_v + 1)$. Then the transition rules (3) − (6) can be used throughout. The decisive precondition for such a system with periodic boundary conditions is the choice of a sufficient size of the system. There must be enough space for the development of the typical spatial structures.

Now our model is complete. The temporal evolution is determined by the transition rules (1) − (6). That means if the phases $Z(i,j)$ of all squares (i,j) at the time t are given, the values of $Z(i,j)$ can be calculated by (1) − (6) one time step later, i.e. at the time $t + 10$ years. This is done by a computer. The variability of the interaction (5) and (6) is taken into account. Thus, rules (5) and (6) are applied only with the corresponding probabilities P_s, P_w and P_e using a random generator. The spatial structure of $Z(i,j)$ can be illustrated as in Fig. 2. Other possibilities of its evaluation will be shown later.

This iterative procedure for the determination of the temporal evolution must start at a certain time $t = 0$. Here, we must give the initial spatial structure $Z_0(i,j)$. We choose a random distribution of the 55 phases over the patches of the system (see Fig. 2), because we want to investigate whether a spatial structure arises in the system itself.

In order to assess the size of the whole system and the magnitude of possible spatial patterns, we must indicate the size of the squares. For this purpose we consider the solar radiation effect. In our model we assume that a total square of beech trees will change to an opening within 10 years, if in the southern neighbourhood an opening exists. Only the trunks of the first trees situated immediately at the border of the opening are exposed to the sun. These trees die within 5–8 years (Nicolai 1986; Remmert 1987). Thus, a strip which is almost twice the diameter of the width of a beech tree can die due to solar radiation in 10 years. Since the usual diameter of a beech crown (*Fagus sylvatica*) is about 20 m, the side of a square is about 30 m. This rough estimation is sufficient since models can only provide the order of magnitudes of the results because of their simplicity and their idealizations.

Finally, some comments to this approach must be given. The local dynamics of the system are not deduced as done predominantly in ecological modelling (Hassel 1978; Christiansen and Fenchel 1977; Wissel 1989; Yodzis 1989). Instead, a semi-phenomenological description is used by imposing the basic local dynamics (1) and (2), i.e. the periodic cycle. The causes for this cycle are not modelled.

The dynamics are modified by the interaction of the squares. More emphasis is placed on modelling the spatial structure by Z(i,j) and its global dynamics.

3 Results

3.1 Main Properties

In the following the results of the model are analyzed. Quantities must be considered which can answer the questions of interest. Above (Sect. 2.1) we asked whether cyclic behaviour can also be seen in quantities which describe the whole system. A quantity of this type is represented by the total number of squares N_b which contain beech trees. In Fig. 4 this number N_b is shown versus time t. Clear oscillations appear with a periodic time of T = 440 years. That this T is shorter than the original cycle length of 550 years is understandable since the neighbour interactions (3) − (6) result in a short-cut of this normal cycle. The maximal amplitudes of the oscillation in Fig. 4 are N_b = 600 and 300, respectively, i.e. the portion of the 1008 squares which contain beech fluctuates between 30 and 60%.

The number of openings N_o shows oscillations of the same periodic time (see Fig. 5). Here, the relative amplitudes are larger. Because of the smaller total

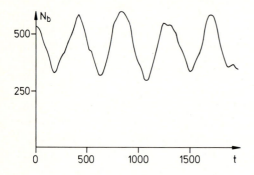

Fig. 4. Total number N_b of beech squares versus time t in years for a system of 1008 squares. (P_o = 0.2; P_s = 0.5; P_w = 0.7; P_e = 0.1)

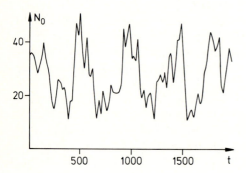

Fig. 5. Total Number N_o of openings versus time t in years

number of openings, the random property of the neighbour interactions (5) and (6), described by the probabilities P_s, P_w and P_e, produces a stronger effect resulting in the random fluctuations superimposed upon the regular oscillations. As expected, the maximal number of openings appears when the number of dying beech trees is maximal. This is the case at t = 500 and t = 1000, for instance when the number of beech trees decreases. Oscillations of the same periodic time with the corresponding phase shift can also be seen for the number of squares with birch trees and mixed forest.

The next question of interest is whether the spatial pattern remains random as at the beginning (see Fig. 2) or whether a spatial order arises. In Fig. 6 the states of the squares are shown at time t = 1710 years, i.e. in a period with a decreasing number of beech trees. It is obvious that a non-random spatial pattern appears. There are patches of several squares which contain beech trees (spotted to black) only, or birch trees (striped) or mixed wood (grey). There are squares with openings (white) scattered in between. Remember that the age of the beech trees is indicated by the blackness of the squares, i.e. the lightly spotted squares indicate young beech trees, the black ones old trees. It is obvious that Fig. 6 shows patches of trees of old (middle), young (lower right side) and moderate age (lower left side). Thus, we find patches of forest with a single age class. The typical

Fig. 6. Spatial pattern of a system of 1008 squares when the number N_b of beech squares decreases. See Fig. 3 for explanation of squares. North is at the *top*

length scale of the patches is three to five side lengths of the squares, i.e. 90–150 m.

It would be desirable to compare these theoretical results with empirical data. However, no virgin forests exist in Middle Europe. But there are data from a Yugoslavian virgin forest (Mayer et al. 1980) which show a spatial mosaic with patches of the same length scale (100–150 m) as in our model.

In order to obtain a better understanding of the results, several parts of the model have been changed. The early colonizations by birch (3) and by mixed wood (4) have been omitted without essentially changing the results. Only the periodic time T is slightly enlarged by this change in the model. Thus, the factors (3) and (4) may be realistic, but they are not essential for the results of the model. The same is true for the dying probability P_0 of old beech trees. If one chooses $P_0 = 0$ or $P_0 = 1$ or values between 0 and 1, there is no substantial change in the results.

The decisive factor of the model is the neighbour interaction (5) and (6), which describes the effect of solar radiation from a neighbouring opening. If this interaction is dropped (i.e. $P_s = P_w = P_e = 0$), no spatial pattern is formed. It has to be emphasized that it is very decisive that a certain variability of this interaction is included. If one assumes that this acts deterministically ($P_s = P_w = P_e = 1$), very strange geometric patterns will arise which are completely unrealistic. Thus, there must be a finite probability ($0 < P_s, P_w, P_e < 1$) for it. This variability is surely present in reality. There is great variability in the reaction of the trees when their trunks, depending on their shape, are exposed to solar radiation. Solar radiation fluctuates with the weather conditions. The stand may be dryer or wetter, and there may be a varying impact of pests. In nature many different factors exist which may influence this interaction.

The concrete values of P_s, P_w and P_e between 0 and 1 are not so important, although several combinations have been investigated. They influence the periodic time T and the amplitudes to a small degree. But the qualitative results are independent of the special values, and the orders of magnitude of the resulting quantities also remain unchanged. Therefore, the values $P_s = 0.5$, $P_w = 0.7$ and $P_e = 0.1$ are used throughout in the following which may be realistic (Nicolai 1989, pers. commun.).

In order to obtain an understanding why this variable interaction (5) and (6) is decisive, one can argue in the following way. The neighbouring interaction (5) and (6), describing the effect of solar radiation, causes a synchronization: An opening causes another opening in the northern direction in the next time step presuming that beech trees are present. The variability of this interaction, described by the finite probabilities P_s, P_w and P_e, weakens this synchronization. Therefore, patches of finite size only are created. A reason for the oscillations of the total number of beech trees (see Fig. 4) will be discussed in Section 3.3.

3.2 Detailed Analysis

It is rather difficult to give an exact measure for the sizes of the patches. In Fig. 6 it can be seen that an exact delimination of a single patch is not possible. A possible quantification of the length scales of the patches can be achieved in the following way: Going through the lines, one counts the number of connected beech squares. For instance in the first line, from left to right, there is a single square ($l = 1$), then $l = 7$ and finally $l = 4$. Continuing in this way in the other lines and columns, one can determine the frequency distribution of different patch lengths l. The corresponding histogram is shown in Fig. 7. In Fig. 8 a histogram, resulting from a random distribution as in Fig. 2, is also shown for reasons of comparison. In Fig. 7 the frequency is shifted strongly to higher values of l, indicating the larger patches. However, a large number of small ones remains. In the following we will consider the mean values $M(l)$ and the standard deviations $Sd(l)$ of such histograms.

Another possibility to characterize the state of the whole system, disregarding the spatial distribution, is the determination of the frequency distribution of the different values Z for the phases. Remember that Z describes which species of trees (birch, mixed wood, beech) is present in a square and represents the age of these trees. Figure 9 shows the histogram for this frequency when the number of beech trees in the whole system increases. Beech trees are described by $Z > 22$. Thus, Fig. 9 shows that the age distribution of the beech trees is shifted to the young trees. This is typical for a growing population (here, beech trees). Below we will consider the mean value $M(a)$ of the age of the beech trees.

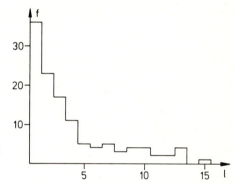

Fig. 7. Frequency (f) distribution of patch lengths l (a unit is the side length of a square) when the number N_b of beech squares is minimal

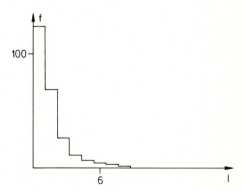

Fig. 8. Frequency (f) distribution of patch lengths l for a random, spatial distribution (see Fig. 2)

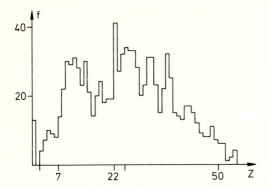

Fig. 9. Frequency (*f*) distribution of the phases *Z*, i.e. of the age of the trees when the number N_b of beech squares increases (Z > 2 birch; Z > 7 mixed forest; Z > 22 beech)

In order to investigate the temporal-spatial properties of our model, we consider the temporal variation of the mean patch length M(l). In Fig. 10 we see that M(l) shows the same strong oscillation as seen in Fig. 4 for the total number of beech trees N_b. Both oscillations are in-phase, i.e. the largest patches appear when the number of beech trees is maximal. Figure 11 shows strong oscillations of the standard deviation Sd(l) of these patch lengths. Their maxima coincide with the maxima of the mean value M(l) in Fig. 10, i.e. in the presence of the largest patch sizes, small patches also appear. Therefore, the standard deviation Sd(l) is large in this situation.

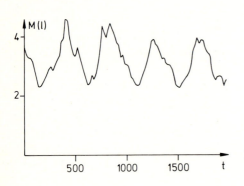

Fig.10. Mean patch length *M(l)* versus time *t* (a unit is the side length of a square)

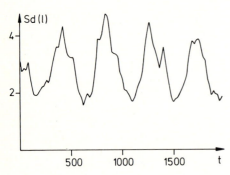

Fig. 11. Standard deviation *Sd(l)* of the patch length versus time *t*

Next, we consider the temporal variation of the mean age M(a) of the beech trees (see also Fig. 9). Figure 12 shows strong oscillations of M(a) in the course of time. They are synchronous to the oscillations of the total number N_b of beech trees in Fig. 4, i.e. they show the same periodic time T = 440 years. There is a phase shift between Fig. 4 and Fig. 12, which means that the mean age M(a) is maximal where the number N_b of beech trees decreases (at 500, 950, 1400) and is minimal where N_b increases (at 300, 750, 1200). This typical shift of the age distribution to young or old individuals respectively for an increasing or decreasing population has been shown already in Fig. 9.

Finally, the temporal variation shown in Figs. 10–12 is illustrated by the spatial pattern shown in Fig. 13. As discussed above, many lightly dotted squares can be seen in Fig. 13a, i.e. young beech trees indicate an increasing beech population. This is in contrast to Fig. 6 where many dark squares, i.e. old beech trees, are seen, thus indicating a decreasing beech population. In Fig. 13b the spatial pattern is shown for maximal numbers of beech trees. Very large patches can be seen, but there are also small ones. This agrees with the variation of the mean patch size M(l) and its standard deviation Sd(l) as shown in Figs. 10 and 11. In Fig. 13b areas with young (lightly spotted) and old trees (dark) are indicated. Generally, patches consisting of beech forest of the same age class are shown by the blackness of the squares. This can also be seen in Fig. 13c where the number of beech trees is minimal.

3.3 System Size Dependence

The investigation of the influence of the system size on the results discussed above will provide more insight into the functioning of the system. In Fig. 14 the total number N_b of beech trees is shown versus time t for a larger system size of 9144 squares. (The results shown in Fig. 4 correspond to 1008 squares.) Oscillations of the same periodic time as in Fig. 4 can be seen. At the beginning, the relative amplitudes are also the same. But later, a damping occurs in Fig. 14 and finally only small oscillations remain. There is a general tendency that the smaller the system, the larger the amplitudes of the oscillations and the more random

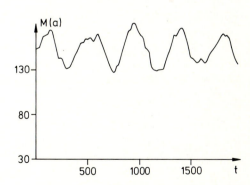

Fig. 12. Mean age *M(a)* of the beech trees versus time *t*

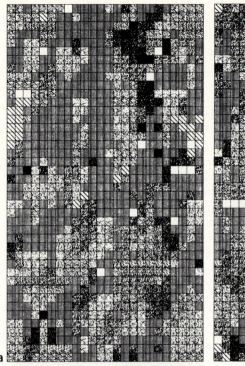

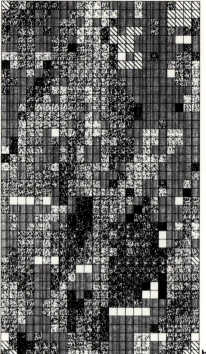

Fig. 13. Spatial pattern when the number N_b of beech-squares is **a** increasing; **b** maximal; **c** minimal (decreasing, see Fig. 6). See Fig. 3 for explanation of the squares

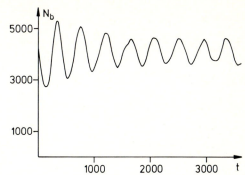

Fig. 14. Total number N_b of beech squares versus time t for a system of 9144 squares (cf. Fig. 4)

fluctuations are superimposed. For very small systems (number of squares less than 100) these random fluctuations are so strong that no regular oscillation can be detected in between. This randomness results from the varying neighbour interaction (5) and (6) described by the probabilities P_s, P_w and P_e in our model. It is well known that the law of great numbers results in a smoothing of this random variation. Thus, the smoothness of the oscillations in Fig. 14 is understandable.

In order to obtain an understanding of the magnitude of the amplitudes of these oscillations we subdivide the rectangular system into six subsystems as shown in Fig. 15. The number of beech trees in the six subsystems and in the total system is determined in the course of time and shown in Fig. 16. In Fig. 16a the usual oscillations for the total system with 1008 squares are seen (see also Fig. 4). In Fig. 16b the numbers of beech trees in subsystems 2 and 3 are shown versus time. As expected from the discussion above, the oscillations in these smaller subsystems are stronger, and random fluctuations are superimposed. But the oscillations in the subsystems are out of phase between each other and out of phase with the oscillation of the whole system in Fig. 16a. In addition, the periodic times in subsystems 2 and 3 are slightly different.

Regarding these results, the outcome in Fig. 16a becomes understandable. If we add the numbers of beech trees of the subsystems, the phase shifts cause a partial cancellation of the oscillations, resulting in an oscillation of smaller amplitudes for the whole system in Fig. 16a. When the system size is large, more

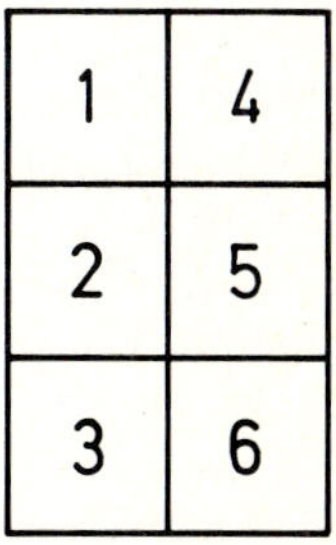

Fig. 15. Subdivision of the system in six subsystems

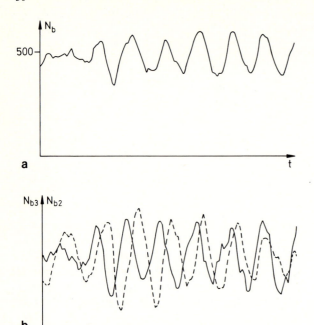

Fig. 16. Number N_b of beech squares versus time t for a system of 1008 squares: **a** for the whole system; **b** for subsystem 3 (*continuous line*) and subsystem 2 (*dotted line*)

subsystems with different phases are included and the amplitudes of the resulting oscillation should be smaller (see Fig. 14). Finally, we expect that the total number of beech trees of a very large system does not show any oscillation, but remains constant.

The same arguments can be used to understand another phenomenon, i.e. the long time variation of the number N_b of beech trees (Fig. 17). There are conspicuous periods in which no oscillations can be detected. Let us remember that the periodic times of the subsystems are slightly different (see Fig. 16b). Thus, the phase relations of the different subsystems change slowly in time. Therefore, we expect that there may be periods in which most of the subsystems are in-phase and their summation results in a pronounced oscillation (at $t = 4000 - 5000$, around $t = 12\ 000$). In other periods, phase shifts of a different degree appear, resulting in smaller oscillations. Occasionally, the subsystems may be totally out of phase. In this case the oscillations cancel each other completely and random fluctuations can only be seen in the total system (at $t = 10\ 000$ and at $t = 14\ 000$).

The effect of the system boundary has been investigated using either periodic boundary conditions or boundaries without the neighbour interaction (5) and (6) as discussed in Section 2. For larger system sizes (number of squares greater than 100) no differences can be detected using these different boundary conditions. For systems with less than 100 squares the random fluctuations are so strong in both

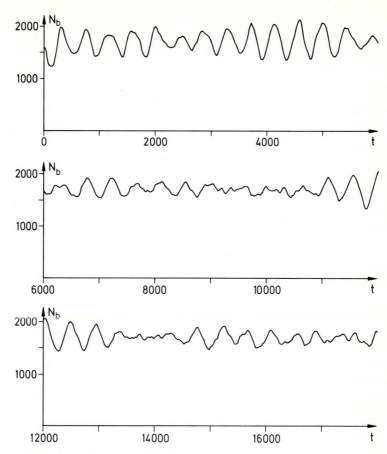

Fig. 17. Temporal (t) variation of the number N_b of beech

of these cases that no regular behaviour can be seen and compared. Thus, the boundary conditions do not play any role for our purpose.

Now we can give a comprehensive interpretation of the behaviour of the model. Essential ingredients of the model are the local cycles, the neighbour interaction (5) and (6) caused by solar radiation and the variability of this interaction. The neighbour interaction causes a synchronization of the cycles in neighbouring squares. Because of the variability of this interaction, this synchronizing effect cannot act over a great distance. Patches of finite size only arise in which the cycles are strongly synchronized. These are the patches with beech trees of the same age for instance (see Figs. 6, 13). The larger the areas under consideration, the weaker the synchronizing effect of this neighbour interaction because of the variability of this interaction. This weakening of the synchronization over long distances can be seen for instance by the reduction of the relative amplitudes of the oscillations in large areas (see Figs. 4, 14).

Thus, three spatial scales exist in our system. The smallest is the size of a single square. In Section 2 we argued that the side length should be about 30 m. It is determined by the distance over which the neighbour interaction of solar radiation can act in 10 years. The second spatial scale is determined by the size of the beech patches. Their typical diameter is 90 to 150 m. The third length scale is the correlation length which determines how far the synchronizing effect of the neighbour action reaches. For areas with a diameter essentially greater than this correlation length, no oscillations of the total number of beech trees can be seen. The order of magnitude of this correlation length is 4000 squares, i.e. about 12 000 m.

4 Stability

The results of our model are suitable for better understanding of the meaning of ecological stability. The definitions of ecological stability are often vague and inaccurate. Several different definitions are possible (Pimm 1982; May 1973; Wissel 1989; Yodzis 1989). For our model we chose constancy as the measure of stability, i.e. the system is considered to be more stable, when the temporal variations of the quantities which describe the state of the system are smaller.

Let us take the number of beech trees N_b to describe the state of the system. We have seen above that the spatial scale is decisive for the temporal variation of the number N_b of beech trees. A single square changes permanently its state because of the cycle, which means on a spatial scale of about 20 m there is no constancy. In this case the system appears unstable. But if we consider an area much larger than the correlation length of 12 000 m, the total number N_b of beech trees is constant. Thus, a system of this size appears to be stable with respect to the quantity N_b. We have seen that the oscillations of N_b are stronger, when the area under consideration is smaller. Therefore, we can conclude that smaller systems are less stable.

But also the time scale under consideration is important for the assessment of ecological stability. On an evolutionary time scale stability does not exist. Our considerations above are valid on a time scale of some 100 years. But for a shorter time scale, e.g. 5 years, a patch of beech trees remains constant. Therefore, this patch may offer a very stable environment for a short-lived animal with a generation time of a few years. Only if we consider a population of this species over many generation times does this patch appear to be ecologically unstable. On the other hand, let us consider an animal which moves around over long distances during its daily life. In this case it covers a large area of the system. The system which it experiences during the day is constant, i.e. stable. Thus, the temporal and spatial scale considered here is essential for the assessment of ecological stability. A survey is given in Table 1.

Another aspect of ecological stability is the reaction of the system to different disturbances. The forest may be used for timber trade. Therefore, we consider the case in which all beech trees older than 140 years are cut. Figure 18 shows the reaction of the system in which the cutting takes place when N_b is minimal

Table 1. Ecological stability of the model system depending on the spatial and temporal scales

Time scale	Spatial scale	
	Small	Large
Small	Stable	Stable
Large	Unstable	Stable

during the oscillation. Qualitatively there is no difference before or after the disturbance at the time t′; the number N_b oscillates in both cases. But the magnitudes of the amplitudes are considerably enlarged by the cutting, indicating a destabilization. These stronger oscillations are damped in the course of time.

Another quantity used for studying the reaction of the system is the mean patch length M(l). Its dramatic reaction is shown in Fig. 19. The first reaction is a small reduction in the sizes of the beech patches due to the cutting. But then extreme oscillations of the patch sizes occur. The minima remain the same as before but the maxima indicate extremely large patches, which arise as many squares have openings due to the cutting. This overload of openings results in a high number of squares in the initial phase of the cycle. They come simulta-

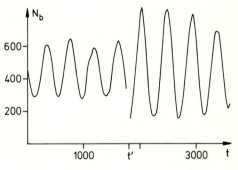

Fig. 18. Reaction of the number N_b of beech squares to the cutting of all beech trees older than 140 years at time t′ (at minimal N_b during the oscillation). Total number of squares is 1008

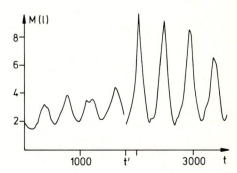

Fig. 19. Reaction of the mean patch length M(l) to the same disturbance as in Fig. 18

neously into the beech phase later, giving rise to the large number of beech trees
with very large patches.

A reaction to the disturbance can also be seen at the mean age M(a) of the
beech trees. In Fig. 20 it is shown that the cutting of the older beech trees results
first in a reduction of the mean age M(a). We argued above that a great number
of squares run in-phase through the cycle. Thus, we expect to find later a large
number of old beech trees representing a high mean age M(a) as seen in Fig. 20.
These great numbers of squares run farther through the cycle in-phase giving rise
to these strong oscillations. Thus, the disturbance results in a synchronization of
large spatial scale. The stochasticity of the neighbour interaction will destroy the
strong phase relation of these many squares in the course of time. Therefore, the
damping of these strong oscillations appears. But it takes several 1000 years until
a state with the original magnitude of the amplitudes is reached.

Next, we consider the situation in which the same disturbance (cutting of
beech trees older than 140 years) is given at a moment t′ where the oscillations
show the maximal number of beech trees. The corresponding reaction is shown
in Fig. 21. Before and after the time t′ oscillations are seen which are typical for
an undisturbed situation (see Fig. 17). The disturbance at t′ reduces the number
of beech trees down to a value which would also be reached by the oscillation in
the course of time. The oscillations simply continue from this value. Thus, the only
result of the disturbance is a phase jump at t′. This type of reaction is expected in

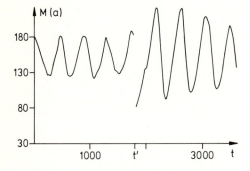

Fig. 20. Reaction of the mean age *M(a)* of
beech trees to the same disturbance as in
Fig. 18

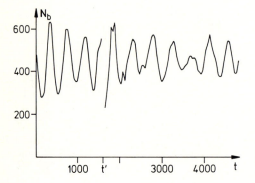

Fig. 21. Reaction of the number N_b of
beech squares to the same disturbances as
in Fig. 18, but now at the moment *t′* of
the maximal number N_b during the oscillation. Total number of squares is 1008

all cases in which the disturbance sets the system into a state which would be reached by the oscillations without the disturbance in the course of time.

So far we considered a system with 1008 squares. Now we investigate the same type of disturbance (cutting of beech trees older than 140 years) in a system with 9144 squares. In Fig. 22 it is shown that almost the same amount of reduction in the number N_b of beech trees (50%) as in the smaller system (Fig. 21) is caused by the disturbance. But here the resulting state shows a low number N_b which is never reached by the oscillations in the undisturbed system (see Fig. 14). In this larger system the undisturbed oscillations are lower as discussed above. Therefore, the disturbance results in a well pronounced reaction, i.e. in larger oscillations which are damped to the original size in about 2000 years.

One can also use other quantities for the investigation of the reaction to the same disturbance. A very strong reaction can be seen for the number N_0 of openings in Fig. 23. It is evident that first a larger number of openings are created by cutting. But shortly afterwards this number decreases to a value which is near zero. This can be understood as all old beech trees are cut. The larger number of openings is reduced during the next 20 years by the colonization of birch and mixed wood. But then new openings are not created as normally accomplished by the falling of old beech trees, since these old trees have been cut.

This lack of openings after the disturbance presents serious consequences for the system: usually there is a large variety of species specialized on openings. In the undisturbed system an opening exists for 10 to 20 years only. But if an opening

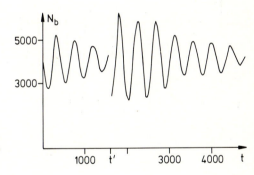

Fig. 22. Reaction of the number N_b of beech squares to the same disturbance as in Fig. 21, but now for a system with 9144 squares

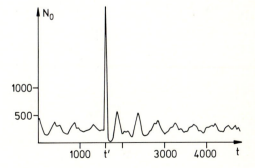

Fig. 23. Reaction of the number of openings N_b to the same disturbance as in Fig. 22.

disappears, other openings are created in the neighbourhood, which have to be colonized by these species. It is a normal process that local populations disappear together with the openings but others are created by the colonization of new arising openings. In this way the survival of a species is guaranteed in the whole system. This procedure is interrupted by the disturbance. For some years only an extremely small number of openings exist. These may be insufficient for the colonization of the larger number of openings which appear later on (see Fig. 23). Thus, the disturbance may interrupt the normal sequence of colonizations. In this case the existence of those species is endangered if a colonization cannot take place from outside the system.

Finally, we apply a stronger disturbance in our model system: All adult beech trees are cut at the time t'. In Fig. 24 the reaction of the number N_b of beech trees is shown. It is evident that at t' the number N_b is set to zero by the disturbance. As discussed above this results in a synchronization of a large spatial scale which is now more extreme than above. A high number of squares run in-phase through the cycle resulting in the extreme oscillations seen in Fig. 24. The number N_b of beech trees oscillates between 0 and 1000, i.e. either no beech tree is present or all squares are occupied by the trees. In this case the system appears rather homogeneous and no spatial structure with beech patches can be seen.

In Fig. 25 the corresponding mean age $M(a)$ of the beech trees is shown versus time t. Only a young beech thicket (less than 30 years old) remains due to the

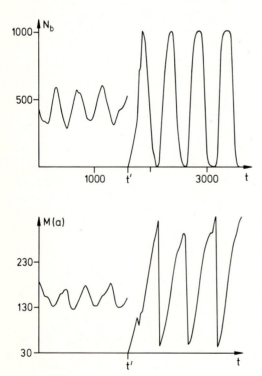

Fig. 25. Reaction of the mean age *M(a)* of beech trees to the cutting of all beech trees

cutting of the adult trees at t'. Then an extreme oscillation is seen between very old or very young beech trees. In contrast to the weaker disturbance in Figs. 18–23, the stronger oscillations are not damped to lower amplitudes here, resulting in a qualitative difference to the above case. In this case the disturbance is so strong that finally the whole system runs through the cycle in-phase. The standard deviation (Fig. 26) of the age of the beech trees decreases on the average in the course of time, i.e. the synchronization of the squares gets stronger and stronger. Finally, the whole system oscillates entirely as a big patch with trees of the same age. In this case no spatial pattern remains and the situation is totally homogeneous in space.

It is not necessary to emphasize that the consequences are catastrophic for the ecosystem. There are no openings for more than 100 years. Thus, species dependent on them cannot survive. This species deficiency is typical for man-made monostands.

We have seen above that the reaction of the system to similar disturbances can be quite different. The timing of the disturbance is very essential. At certain instances of time there will be no reaction of the system. The size of the perturbed area is important. The smaller the area is, the stronger are the natural oscillations before the disturbance and the smaller is the difference of the reaction to these normal fluctuations. The assessment of the system's reaction may be different depending on the quantity which is investigated. By increasing the strength of the disturbance, the reaction of the system will change abruptly at a certain level. After smaller perturbations the system returns to the original states. If the strength of the disturbance exceeds a critical level, the system does not return but changes into a completely different state with a different temporal-spatial pattern. Once this homogeneous state is reached the system will never return to the mosaic structure by itself.

5 Conclusion

We can draw several conclusion from our model results. The aim of our model was to investigate the consequence of the mosaic-cycle concept on the temporal and spatial pattern.

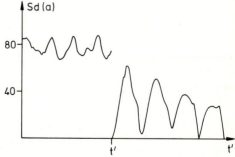

Fig. 26. Reaction of the standard deviation *Sd(a)* of the age of the beech trees to the cutting of all beech trees

Our approach differs completely from the usual methods used in ecological modelling. We do not model the mechanisms which cause the local dynamics, i.e. the cycle, which is usually the case (Hassell 1978; Christiansen and Fenchel 1977; Wissel 1989; Yodzis 1989). We simply adopt it from empirical experience and start with the cycle as shown in Fig. 1. More emphasis is placed on modelling the global dynamics and the spatial pattern of the whole system. The investigations of spatial patterns of ecological models usually rest on mathematically difficult equations, e.g. reaction-diffusion equations (Murray 1989), or use highly complex computer models. Neither of these approaches provide a good understanding of the results. Our approach uses a type of cellular automation with very simple rules. In this way it is possible to obtain good insight in the functioning of the system.

Essential ingredients of our model are the local dynamics, i.e. the cycle in the squares, the neighbour interaction caused by the solar radiation effect and the variability of this interaction. This interaction causes a synchronization of the cycles in neighbouring squares, resulting in the formation of patches with trees of the same age. In larger areas oscillations of several quantities can be seen which are caused by this synchronization. The variability of this interaction weakens this synchronizing effect especially over long distances. Thus, in a large system the synchronization acts only in smaller subsystems causing an oscillation. However, these oscillations of the subsystems are out of phase. Therefore, in very large areas with several subsystems of this type the oscillations of these subsystems cancel each other and only small oscillations remain. Their amplitudes are smaller when more subsystems are included, i.e. when the whole area under consideration is larger. Oscillations in very large systems are not expected.

The word "stability" is often very vaguely used in ecology. Our model shows a way of assessing ecological stability. It is shown that the spatial and temporal scales under consideration and the quantity which is used for the measurement of the reaction are decisive. If one considers the disturbance of the system, e.g. the cutting of beech trees, the timing of the cut, the size of the perturbed area and the amount of cutting are important. It should be mentioned that the system does not show any essential reaction if the disturbance causes a state which would be reached by the system alone in the course of time. Some perturbations result in a state of the system which is included naturally in the cycle (Remmert 1987). In this case the system simply continues in this cycle starting from this state.

Finally, it is desirable to generalize the results of our model for other ecosystems. We expect to find systems with pronounced mosaic cycle behaviour if the following conditions are fulfilled. There must be a local cycle. The normal life cycle of one species may be sufficient if this is the key species which forms the ecosystem (Mueller-Dombois 1987, 1988). In addition, there must be a synchronizing factor which sets neighbouring areas in the same phase of the cycle. Many different synchronizing factors are possible (Mueller-Dombois 1987, 1988). In our model this factor (solar radiation) acts continuously. However, if the main synchronizing factor appears very seldom, then the synchronized areas run in-phase through the cycle as long as desynchronizing, random effects are small enough. It is obvious that stochastic factors (e.g. different lifetimes of the individuals) are desynchronizing, acting in opposition to the synchronizing factors. The relative strength and the spatial action of these two factors determine the resulting mosaic-cycle pattern.

Acknowledgements. The author is very grateful to H. Remmert who initiated this work and supported it by many useful discussions. This work is supported by the Deutsche Forschungsgemeinschaft (DFG).

References

Burton PJ, Mueller-Dombois D (1984) Response of *Metrosideros polymorpha* seedlings to experimental canopy opening. Ecology 65:779–791

Christiansen FB, Fenchel TM (1977) Theories of populations in biological communities. Springer, Berlin Heidelberg New York

Hallam TH, Levin SA (eds) (1986) Mathematical ecology, an introduction. Springer, Berlin Heidelberg New York Tokyo

Hassell MP (1978) The dynamics of arthropod predator-prey systems. Princeton Univ Press, Princeton, New Jersey

May RM (1973) Stability and complexity in model ecosystems. Princeton Univ Press, Princeton, New Jersey

Mayer H, Neumann M, Sommer HG (1980) Bestandsaufbau und Verjüngungsdynamik unter dem Einfluß natürlicher Wilddichten im kroatischen Urwaldreservat Corkova Uvata, Schweiz. Z Forstwis 131:45–70

Mayer H, Neumann M (1981) Struktureller und entwicklungsdynamischer Vergleich der Fichten-Tannen-Buchen-Urwälder Rothwald/Niederösterreich und Corkova Uvala/Kroatien. Forstwiss Zentralbl 100:111–132

Mueller-Dombois D (1987) Natural dieback in forests. BioScience 37:575–583

Mueller-Dombois D (1988) Community organization and ecosystem theory. Can J Bot 66:2620–2625

Murray JD (1989) Mathematical biology. Springer, Berlin Heidelberg New York Tokyo

Nicolai V (1986) The bark of trees: thermal properties, microclimate and fauna. Oecologia 69:148–160

Pimm SL (1982) Food webs. Chapman and Hall, London

Putman RJ, Wratten SD (1984) Principles of ecology, Croom Helm, London

Remmert H (1985) Was geschieht im Klimax-Stadium? Naturwiss 72:505–512

Remmert H (1987) Sukzessionen im Klimax-System. Verh GfÖ Bd XVI:27–34

Sprugel DG (1976) Dynamic structure of wave-regenerated *Abies balsamea* forests in northeastern United States. J Ecol 64:889–911

Wissel C (1989) Theoretische Ökologie: eine Einführung. Springer, Berlin Heidelberg New York Tokyo

Wolfram S (1986) Theory and applications of cellular automata. World Sci Publ, Singapore

Yeaton RI (1978) A cyclic relationship between *Larrea tridentata* and *Opuntia ceptocaulis* in the northern Chihuahuan Desert. J Ecol 66:651–656

Yodzis P (1989) Introduction to theoretical ecology. Harper and Row, New York

The Mosaic Theory and the Spatial Dynamics of Natural Dieback and Regeneration in Pacific Forests

D. Mueller-Dombois

1 Introduction

Most patterns of vegetation, particularly when projected on a map, form a mosaic. But mapped vegetation patterns are generally considered to be static, particularly if they are well correlated with strong differences in habitat.

The mosaic theory, more correctly called the "shifting mosaic" theory or the "mosaic-cycle" concept, refers to a dynamically shifting mosaic that may be detected in a broader community matrix, such as a formation or biome. The theory probably goes back to Aubreville (1938), who observed in the tropical rain forests of West Africa that the saplings or juvenile trees of the canopy species are usually not found under the canopy trees themselves, but spatially displaced in separate areas where former canopy trees had collapsed. Jones (1945) noted similar spatial displacements of regeneration patterns in unmanaged northern temperate forests of Europe. Bormann and Likens (1981) found the same principle applicable to temperate forests in the northeastern United States. These authors referred to it as the "shifting mosaic steady state" which is initiated by tree-fall gaps. The term "gap dynamics" or "patch dynamics" is also used to denote the same phenomenon. The latter term was preferred by Pickett and White (1985) to free the concept from the steady state or climax concept, with which it is sometimes said to be in conflict (Sprugel 1976; Veblen 1985). It is indeed in conflict, if a climax forest is interpreted as having a permanently stable canopy.

Patterns of plant death and regeneration are observed as moving laterally through the plant community. Watt (1947), who introduced the "pattern and process" terminology, theorized that a pattern of spatial dynamics is applicable to all life-form communities or formations such as grassland, heath, scrub, and forest. European forest scientists found broader stand-level patterns, such as multiple tree falls of at least 0.5 ha, and similarly sized regeneration patches, both overlapping and in separate locations throughout several remnant areas of European virgin forests (Zukrigl et al. 1963; Leibundgut 1959, 1978; Mayer and Neumann 1981). They produced large-scale structural-dynamic maps, referred to as displaying "*Waldtextur*" (Mayer et al. 1980, also in Mueller-Dombois and Ellenberg 1974, p. 398 and Mueller-Dombois 1987, p. 576).

Department of Botany, University of Hawaii, Honolulu/Hawaii, USA

H. Remmert (Ed.)
Ecological Studies Vol. 85
© Springer-Verlag Berlin Heidelberg 1991

Dieback is a structural pattern of canopy breakdown whereby most of the trees die for no obvious reason and remain standing upright. They later collapse from fungal disintegration or get toppled by wind. Spatially, dieback is a stand-level phenomenon involving groups of dying trees in various configurations, from less than 1 ha to over hundreds of hectares.

In the following, I will discuss the spatial dynamics of some Pacific forests in which stand-level dieback has been shown to be a natural and recurring phenomenon. This will also include their regeneration patterns and a consideration of the equilibrium concept. Then I will discuss the disturbance regimes and processes that may lead to dieback, and conclude with a general and testable hypothesis that explains why stand-level dieback occurs in some natural forests but not in others.

2 Spatial Patterns of Natural Dieback and Regeneration in Pacific Forests: Four Examples

Although stand-level dieback has been observed and researched in a number of Pacific forests, including, for example, *Tsuga mertensiana* forests in the Pacific Northwest (Matson and Boone 1984; Matson and Waring 1984), *Beilschmidia tawa* and *Ixerba brexioides* forests in the Kaimai mountains of northern New Zealand (Jane and Green 1983), several eucalypt forests in Australia (Walker et al. 1983; Palzer 1983; White 1986; Davison 1988, a.o.), *Nothofagus* forests on mountains of Papua New Guinea (Arentz 1983, 1988; Ash 1988), and Norfolk pine (*Araucaria heterophylla*) dieback on Norfolk Island (Holloway 1977), I will focus here on four examples where the spatial patterns of dieback and regeneration and their shifting dynamics are somewhat better known. The question of patch sizes in the dieback mosaic, and their configurations and spatial shifting over time, requires a landscape-level approach (Mueller-Dombois 1986), which has as yet received little attention in forests with stand dieback.

2.1 The Shimagare Phenomenon in Japan

Kohyama (1988) recently reviewed this peculiar form of stand-level dieback, which occurs in the subalpine fir zone throughout Central Japan on several mountains, extending from the most northern Hokkada to the most southern Ohmine mountains. Its classical spatial pattern of crescent-shaped horizontal stripes, which run parallel to the contours of these mountains at intervals of roughly 100 m, is named after Mt. Shimagare, which in Japanese means "the mountain with the dead stripes of trees". There are two mountain-fir species, *Abies veitchii* and *A. mariesii*, involved in this dieback. The two species are sympatric in the southern half of the mountain range on Honshu Island, while *Abies mariesii* forms mono-dominant stands in the northern half. The striped dieback patterns are restricted to the Pacific side of the mountains and where snow depths are less than 1 m. They are also restricted to an elevational range between approximately 2100 and 2700 m, and are typical for the summit areas of mountains with such heights. On higher mountains, the same fir species grow in the alpine zone, for example, on Mt. Fuji, where such dieback patterns are absent.

In situations where a third tree species, in this case birch (*Betula ermanii*), invades the *Abies* stands during early succession, such dieback patterns are absent. Moreover, in the mixed fir-hardwood stands, the fir species grow to much greater sizes and can become three times as old as in the dieback stands. Kohyama concludes from this that dieback is not a characteristic of the two *Abies* species, but rather is related to their particular stand structure.

Detailed structural and age analyses have shown that the dieback stands are cohort stands, which increase in age and stature in downslope direction toward the regeneration zone, which is directly beneath the dead or dying cohort. The dieback stand can form an 800-m-long band along the contours, varying in width from 20 to 60 m.

Both *Abies* species, in contrast to the *Betula* species, are described as very shade-tolerant climax species. Small seedlings often grow densely under the closed canopy of the living *Abies* stands, but as soon as these stands go into dieback, the seedlings are released from parental competition and grow into saplings. The Shimagare phenomenon is clearly a case of "replacement dieback" or auto-succession. In relation to the shifting mosaic theory, the displacement or lateral movement comes about by the dieback zones moving upslope in parallel stripes at an average speed of 1.3 m/year. The dead standing tree skeletons in the dieback zone eventually break down when the released saplings have gained in stature, representing an early mature age state. The same Shimagare phenomenon has been found on Whiteface Mountain in upstate New York where pure stands of *Abies balsamea* behave in a similar manner (Sprugel 1976).

Sprugel suggested that the Whiteface Mountain fir-regeneration pattern followed Watt's (1947) "pattern and process" concept rather than that of a typical climax forest, while Kohyama (1988) refers to the Shimagare phenomenon as a climax forest in dynamic equilibrium.

2.2 Nothofagus Dieback in New Zealand

Some information on the spatial dynamics of *Nothofagus* dieback at the landscape level is given by Wardle and Allen (1983). The Harper catchment in Canterbury on New Zealand's South Island is covered with a monospecific mountain beech (*Nothofagus solandri* var. *cliffordioides*) forest. Prior to 1968, a closed forest covered an area of 5000 to 6000 ha between an altitude of 650 to 1350 m. This forest was hit by two heavy, moist, snowstorms in 1968 and 1973. Canopy breakage occurred on 35% of the trees. Most of these were in the 20 to 30 cm stem-diameter class. After the first snowstorm a series of 217 permanent, 400 m² plots was established, with 21 000 tagged trees; following the second snowstorm, these were remeasured four times, at 2-year intervals. Thirty percent of the 217 plots showed significant structural damage from the second snowstorm, but there was also continuing mortality that resulted from the first storm in 1968. Eight years after the second storm, mortality was still continuing. But now trees undamaged by either storm were also dying. Thus, there was continuing expansion of dieback among the canopy trees, which was associated with large increases in seedling density. Moreover, the seedlings grew vigorously in height, while the canopy trees continued to die. Some plots (28) were in undamaged stands. The diameters of trees over 40 cm tall, however, did not increase over the remeasurement period and these undamaged survivors suddenly began to die without any obvious cause. Wardle and Allen (1983) remarked that the snow damage synchronized mortality in the canopy cohorts and also encouraged synchronous regeneration.

Another area, the Moa catchment (exceeding 1000 ha), was covered with uniformly sized, monospecific mountain beech forest. Mortality originated in the upper reaches of the catchment and moved progressively downstream over a number of years. The authors suggest that the originating cause in the upper reaches of the valley was wind damage. After that trees died standing upright, and dieback progressed downslope. The forest was apparently in an old-age state. Only a few isolated patches survived, which had originated after more recent disturbances and thus were stands in more juvenile life stages. Similar mass dieback of mountain beech has occurred in Tongariro National Park on the North Island. Skipworth (1983) believed this to be triggered by drought and a lowering of the water table. His structural analyses of eight stands also show that these forests form a mosaic of cohorts in approximately similar age states. Obviously, mountain beech dieback is not a new phenomenon in New Zealand. Skipworth (1983) cites Cockayne, who reported in 1908 that dead standing trees are a "feature" and "that seedlings and saplings are in abundance".

Ogden (1988, p. 226) developed a model for mountain beech dieback in New Zealand that fits the mosaic-cycle theory. He suggested a turnover rate of 120 to 220 years. But it seems clear that an individual catchment covered with mountain beech forest is not in an equilibrium state. It would take a number of catchments of 500 to 1000 ha in size to find juvenile, mature, and senescing age states represented side by side at any given time.

2.3 Scalesia Dieback in the Galápagos

Two recent papers report on the stand-level dieback of the nearly monospecific *Scalesia pedunculata* forest (Itow and Mueller-Dombois 1988; Lawesson 1988). This endemic species, a member of the Compositae, apparently evolved from an ancestral ruderal forb that radiated into at least a dozen different species, most of them became shrubs, while two became definite tree species. *Scalesia pedunculata* is a fast growing tree, resembling secondary tropical pioneer tree species in its rapid growth rate. It reaches a height of 15 m in about 10 years, and has a short life span of only about 15 years (Hamann 1979). The species forms extensive cohort stands on the island of Santa Cruz in the humid zone at middle elevations, a zone that in a continental setting would be occupied by a multispecies tropical moist forest. Only a few other native tree species, such as *Psidium galapageium* and *Zanthoxylum fagara*, are sparingly scattered or only locally codominant throughout the *Scalesia* forest.

Scalesia mass mortality was reported by Kastdalen (1982) to have occurred between 1935 and 1940. Following the most recent 1982/83 giant Niño, which brought extremely heavy rainfall, the entire *Scalesia* forest in the humid zone on Santa Cruz island showed stand-level dieback. Some trees remained standing with much reduced foliage, though they did not recover; most trees collapsed soon after the event. A few years later, the area became restocked with a huge cohort of *Scalesia* seedlings. Some differential growth is showing up in different locations (Itow and Mueller-Dombois 1988), but overall the new growth phase of juveniles exhibits general synchrony in development. The new *Scalesia* stand seems to be setting itself up for another huge dieback, presumably to occur near the year 2000.

The cyclic behavior of this forest has been described also by Hamann (1979). It is a forest that typically displays an autosuccession because it lacks any significant successional tree species that could succeed the pioneer *Scalesia*. Lawesson (1988) reported that fires from human activity have occurred in parts of the *Scalesia* forest zone during the past 50 years and also that cattle and donkeys have run wild since the 1930s. He suggests that these man-induced disturbances have aided in the suppression of successional tree species, favoring further the large-area monodominance of *Scalesia pedunculata*.

The *Scalesia* forest on Santa Cruz is certainly not in equilibrium with its environment, if equilibrium implies a "shifting mosaic steady state". However, the forest has apparently maintained itself through several thousands of years by a simple process of dieback and regeneration.

2.4 The Metrosideros Rain Forest Dieback in Hawaii

From an aerial photo analysis by Petteys et al. (1975), done over an 80 000 ha area covering the eastern slopes of Mauna Kea and Mauna Loa, between 610 and 1830 m altitude on the island of Hawaii, it was seen that dieback began in the 1950s near the NW corner of the sample terrain. In 1954 it covered 120 ha, by 1965 severe dieback had spread over an area of 16 000 ha, and by 1975 it covered 34 500 ha. From this rapid pattern of expansion, the dieback cause was suggested to be a new or introduced epidemic disease. However, the disease hypothesis could not be substantiated (Papp et al. 1979) in spite of several years of intensive disease research.

At first the spatial dynamics of the dieback pattern made little sense. It appeared to represent a random mosaic that coalesced into larger and larger patches. Some patches remained small, less than 1 ha, while others extended over areas of at least 1000 ha. The dieback expansion seemed to continue until about 1977, by which time approximately 50% of a 100 000 ha sample area had gone into dieback (Jacobi et al. 1983).

The dieback and non-dieback mosaics were thoroughly investigated with regard to physical habitat variations and patterns of population structure of

Metrosideros polymorpha (Mueller-Dombois et al. 1980). Several major habitats were distinguished with different structural dieback patterns: (1) an extensive tree-to-tree dieback on poorly drained pāhoehoe lava ("wetland dieback"); (2) a smaller, but multiple-tree dieback, in patches up to 1 ha on 'a'ā lava and other well-drained substrates ("dryland dieback"); (3) a large area of "bog-formation" dieback in extremely poorly drained areas on older soils from volcanic ash; (4) another large-area dieback on moderately well-drained and recent (i.e., 1000 year old) eutrophic ash ("displacement dieback"); (5) a "gap-formation" dieback, in size range similar to (2), but usually with more trees on the ground than standing upright, indicating an older patch dieback.

Thus, in the *Metrosideros* rain forest biome, the major habitat variations in terms of physical gross-textural, soil-nutrient and water-regime, and associated substrate-age differences impose certain restrictions on the shifting mosaic. Although dieback and non-dieback stands occurred on all substrates, there was relatively little dieback in young forests under the age of 100 years. Stand-level dieback was only found among canopy trees. In terms of recovery patterns it was possible to recognize three types:

1. "Replacement dieback", where saplings of the dying canopy species were coming up in some abundance;
2. "Displacement dieback", where the canopy species became successively displaced by other species, typically by tree ferns (*Cibotium* spp.) on the eutrophic (1000-year-old ash) soil, and where only few saplings will form a next generation stand of widely scattered trees; and
3. "Stand-reduction dieback", which is related to an obvious site deterioration for tree growth, in the boggy habitats. Here, the next *Metrosideros* generation will probably only grow to shrub stature.

Dieback in the *Metrosideros* rain forest biome seems to be a one-time event, extending over a period of at least 30 to 50 years. But previous dieback episodes have been reported, and one large-area dieback on the island of Maui, which occurred between approximately 1905 and 1935, still displayed recent dieback in the early 1980s (Holt 1983). This dieback too was first believed to be a spreading disease and thereafter thought to be caused by soil aging associated with iron toxicity (Lyon 1909, 1919). It turned out to be a stand-reduction dieback (Holt 1988), similar in structure to the more recent bog-formation dieback on Mauna Kea.

From the viewpoint of the mosaic theory, the Hawaiian dieback currently does not appear to represent a "shifting mosaic steady state". However, it may be a partially shifting mosaic over a 300- to 400-year time span, which is the approximate generation time for *Metrosideros* trees (Atkinson 1970, Porter 1973). A longer-term shifting mosaic in the *Metrosideros* forest biome, approximating these generation times, is supported by Selling's (1948) analysis of bog pollen and spores (Fig. 1). According to this, *Metrosideros* can be described as a tree species that persisted for at least 10 000 years, but in an oscillating (or shifting) abundance pattern alternating with tree ferns. The current dieback and non-dieback de-

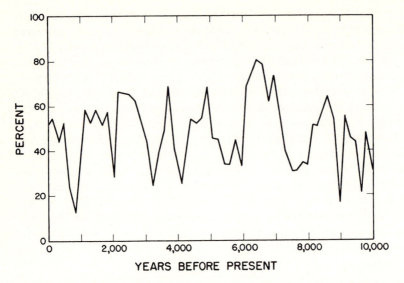

Fig. 1. The percentage abundance of *Metrosideros* pollen found in pollen and tree fern spore counts from a Hawaiian rain forest bog on Moloka'i (after Selling 1948). The time axis is based on the general relationship determined from carbon dating the 325-cm core (Juvik, pers. commun.). Note the fluctuating but persistent presence of *Metrosideros* throughout this period (Mueller-Dombois 1986)

signation is a grossly simplified description. There are a number of younger age states represented on differently aged lava flows and ash deposits. These younger forests are mostly vigorous and without stand-level dieback. Moreover, many of the former dieback stands are now in advanced stages of recovery (Jacobi et al. 1988), i.e., in the "building" phase.

3 Dynamic Processes in Dieback Forests

Inquiries into the causes of the Hawaiian dieback, based on a large number of well-established facts summarized elsewhere (Mueller-Dombois 1985, 1986), led to the realization that at least two cause complexes are involved, one relating to disturbance regimes and the other to stand development.

3.1 Disturbance Regimes

In their synthesis on *Natural Disturbance and Patch Dynamics*, Pickett and White (1985, p. 10) provide a tabulated overview of natural disturbances. Their list is non-hierarchical and includes violent disturbances, such as fire, hurricanes, landslides, and lava flows in the same category as gap dynamics and biotic diseases.

Certainly, gap dynamics is not simply a type of disturbance but rather a theory or concept similar to patch dynamics. Both involve disturbance in the formation of gaps or patches, but the question of how gaps or patches are formed has rarely been asked by ecologists. None of the 14 articles on "tree-fall gaps and forest dynamics" published as a special feature in Ecology (Vol 70, No. 3) address this question.

An answer to the question of how gaps are formed has become very important during the last decade, when forest dieback and decline were noted in forests of Central Europe and eastern North America. An ecological answer was presented in a special feature with seven articles on "tree death" by AIBS (1987). Natural dieback, such as that reported in the above four examples, is certainly a form of gap or patch dynamics.

Such forms of natural forest dieback cannot be explained by a single and obvious cause. At least two kinds of *abiotic* disturbances should be distinguished:

1. Violent or catastrophic stand-destroying disturbances, such as fires, hurricanes, landslides, and lava flows. Among these it is, of course, still important to also distinguish those that initiate a new primary succession (such as lava flows) and those that initiate a secondary succession (such as hurricanes), because these two subcategories differ in so many ways (Mueller-Dombois and Ellenberg 1974).
2. Physiological shocks, such as a transient drought, a severe rainstorm, a sudden drop in temperature, a period of extremely low radiation, or a substrate vibration due to earth tremors. These may temporarily or permanently stress a vigorous forest stand but may not kill it. However, the same disturbance may kill a low vigor stand.

Physiological shock is considered a less devastating category, but shocks may recur several times at intervals of a few years, affecting the same individuals growing together in the same stands and thereby interfering at least temporarily with their normal growth and reproduction processes. Physiological shocks can also be considered as biome-specific pulse perturbations (Mueller-Dombois 1988). If often repeated or occurring with increased intensity, they may bring an individual tree or a cohort stand into premature growth stagnation.

In reality, of course, there is a continuum of abiotic disturbances in terms of their effects, but it seems useful to consider at least the two types, the stand-destroying and physiological shock-creating disturbances. Both types usually occur in the same biome and may affect large or small areas, often related to topography. In the boreal forest (Heinselman 1981) and in southeastern pine stands (Christensen 1981), where fire is a frequently recurring disturbance, cohort stands may rarely grow into an old-age state. But where they are protected from fire they grow naturally into a low vigor state at which time they become vulnerable to insect attack (Schowalter 1985).

Biotic diseases and insect pests can be considered a third type of disturbance. Alien killer diseases, such as white pine blister rust, Dutch elm disease, or chestnut blight, can destroy vigorous stands. In contrast native biotic diseases typically have a regulatory function in natural forests by infecting only the weakened individuals. They usually become aggressive only when the vigor of their natural host tree declines. If the forest system is made up of a mosaic of cohort stands in different age states, they usually attack those stands that are in an old-age state or those that have been forced into stagnation by physiological shocks or other local site stresses. As a disturbance type, biotic agents differ from the two abiotic disturbances, by being dynamically related to the vigor state of t' ɔir hosts. If their hosts occur in cohort stands of physiologically similar indi᾽ .duals in declining vigor, their response will likely be a stand-level infestation. Therefore, indigenous biotic diseases and insect pest species are closely in tune with stand development.

Disturbance events of the abiotic types 1 and 2 are relatively independent of the age/vigor state of cohort stands, while those of type 3, the biotic disturbances, usually depend on the age/vigor state of the host tree or stand. Moreover, the effects of type 1 disturbances are relatively independent of the age/vigor state, while the effects of type 2 and 3 disturbances depend on the age/vigor state of the trees comprising the affected stand.

3.2 Stand Development and Stress

A biome in which stand-destroying disturbances of type 1 occur repeatedly in different locations will favor pioneer species that can take advantage of open areas. This applies to three of the above examples, the Hawaiian, Galápagos, and New Zealand forests. *Metrosideros polymorpha, Scalesia pedunculata,* and *Nothofagus solandri* var. *cliffordioides* are shade-intolerant pioneer species. Seeds of the first two species are frequently produced and wind dispersed. These species colonize open habitats, often forming large cohort stands, which seem to perpetuate themselves over a few generations through stand-level dieback with subsequent replacement. Indications are that subsequent cohort stands become spatially more and more heterogeneous, resulting in smaller-area dieback patterns.

The *Abies* species associated with the Shimagare phenomenon are said to be shade-tolerant climax species adapted to cyclic canopy opening without further clearing of habitat. However, in the harsh subalpine environment, they have a dual function also as pioneer species, since they perform autosuccessions. Moreover, the fir seedlings do not seem to grow into saplings until the parental cohort dies. A clearing of habitat, probably by typhoons, was likely responsible for initially creating the striped patterns of the Shimagare phenomenon. Subsequently, this was maintained by cohort senescence in which endogenous and exogenous stresses, such as physiological shocks and site stresses, acted synergistically to bring about the stand-level dieback.

Stress can be defined as a strain or pressure that follows a disturbance. After successful colonization, stresses will always accompany stand development. In cohort stands, one important stress results from intraspecific competition. Depending on stocking density and the resource base of a habitat, intraspecific competition may begin in the seedling stage, or in later life stages. It continues as long as the trees are actively growing and interacting at least below ground.

Edaphically, extreme habitats impose a second type of stress. Such habitats may not necessarily limit stand establishment if it occurs, for example, during periods of favorable soil-water relations. Edaphic stresses can increase as trees grow in stature. Intraspecific competition by elimination of weaker individuals will be the first expected response, but even the stronger survivors will eventually be weakened and the whole stand can become stunted or forced into a state of "premature senescence". Extreme examples are the so-called toothpick stands frequently occurring, for example, in stands of *Pinus contorta, Picea mariana, Tsuga heterophylla,* and other tree species in northern temperate forests.

A third stress is an inevitable result of life's program itself, which includes death as part of life. Every sexually reproductive species has a potential life span. This life span can be highly synchronized as among a population of annual plants growing in the same habitat. In polycarpic plants, death is generally less synchronized, but a certain synchrony can come about (1) if stand development results in even-structured cohort stands, and (2) through effects of the habitat itself, particularly if it is harsh, implying such limitations as imbalanced nutrients, excessive or poor drainage, or constantly high wind exposure. This added stress will push individuals into a period of stagnation or senescence which predisposes them to die upon an external cue, such as another physiological shock.

The colonization of large open areas by pioneer tree species cannot be expected to be uniform. The invasion pattern depends on the supply and distribution of disseminules, on habitat variations within the opened area, and also on favorable periods for germination and establishment. A pioneer species may colonize a harsh habitat during a favorable period and be forced into dieback when conditions become unfavorable. Such a process of advancement and retreat, followed again by re-advancement coupled with adaptation, is what is understood as "radiation", an important process leading to speciation in islands. Therefore, dieback stands may also be an indication of insufficient site adaptation and therefore, ongoing radiation.

Depending on the relative harshness of the open site and the growth rate of the species, the invasion process may extend over only a few or many years. *Metrosideros* requires at least 30 to 50 years to fully occupy a lava flow in the rain forest terrain about 1 km away from an undisturbed forest. Thus, there will be age gradients within larger cohort stands which are maintained from the colonization stage to maturity. Similar patterns of invasion can be expected in areas devastated by hurricanes, but the re-establishment processes would be much more rapid because of the already developed substrate and the existence of in situ propagules. Depending on biome characteristics, the stand-destroying disturbances may not return to the same locations over several tree generations. In that case, trees will die naturally from a combination of physiological shocks, biotic diseases, and endogenous stresses associated with stand development. It is suggested that stand development is an important determinant of stand-level dieback.

Similar considerations of disturbance regimes and their interactions with stand development and stress resulted in what I have called the "cohort senescence theory" (Mueller-Dombois 1983, 1986), which has been presented as a chain reaction model elsewhere (Mueller-Dombois 1988).

4 Conclusions

In their recent multi-author book on *The Ecology of Natural Disturbance and Patch Dynamics*, edited by Pickett and White (1985), only two kinds of gaps are considered, tree-fall gaps and gaps of various sizes created by stand-destroying single-factor disturbances. Stand-level dieback is not included. The same applies to the recent feature issue on gap dynamics in Ecology (Platt and Strong 1989). Sprugel's (1976) study of fir-wave regeneration on Whiteface Mountain is cited, which is the American equivalent of the "Shimagare phenomenon", but the concern is primarily with what happens after the gaps are formed and very little with why they are formed.

Perhaps this is what motivated Lieberman et al. (1989) to use the phrase "forests are not just Swiss cheese" as the title of their article on forest gap dynamics. At least Whitmore (1989) in his lead article considers the forest growth cycle to consist of three basic phases, the gap, building, and mature phase. He mentions a "degenerating" phase, but dismisses it as unimportant. This viewpoint is understandable as decomposition is usually very rapid in the humid tropics. However, in biomes with slow rates of decomposition, dead standing trees can be expected to be a prominent feature. The "Swiss cheese" analogy criticizes the simplified dichotomy of considering only gaps and mature forests as making up the forest mosaic. Lieberman et al. (1989) believe that the "building phase" needs at least equal attention. Here, I add a fourth phase as deserving equal attention in research. This I have called the "senescing" or "breakdown" phase (Mueller-Dombois 1987). Others refer to it as an "old-growth" phase, while American foresters have used the term "overmature" or "decadent" forest, and population biologists the term "degenerative" phase. If ecological and not just pathological research attention is given to this "stand-aging" phase (to use a European term), studies in forest dynamics may eventually achieve a comprehensive ecological theory that can then address human-related disturbances, such as air pollution of forests, more effectively.

This chapter adds another dimension to the patch dynamics or mosaic theory. Dieback stands are an important structural and dynamic phenomenon in many forested landscapes (Mueller-Dombois and McQueen 1983; Arndt and Mueller-Dombois 1988), and they should be included in any comprehensive theory of forest dynamics.

In terms of their origin, dieback stands take an intermediate position between the primarily endogenously caused single tree-fall gaps typical for the multi-species tropical lowland forests (Hartshorn 1978; Brokaw 1985) and the purely exogenously created gaps caused by clearly identifiable single-factor disturbances. Stand-level diebacks fall into a still more predictable category than the latter, since they are closely connected to stand development.

Moreover, they appear to have a very close relationship to low canopy species diversity. One of the four examples cited was a truly monospecific forest system, the mountain beech (*Nothofagus solandri* var. *cliffordioides*) forest in New Zealand. The *Scalesia pedunculata* forest contains only a few additional canopy species, which in terms of stand dynamics are rather insignificant. The Hawaiian *Metrosideros* rain forest is peculiar due to its near monodominant *Metrosideros* canopy, but it contains a number of lower-statured associated tree species that rarely play an important role as canopy associates except in some successionally advanced forest stands on older soils. While there are many other species and life forms in the *Metrosideros* rain forests on Hawaii (Mueller-Dombois et al. 1981), its canopy is mostly monospecific. The Japanese subalpine fir forest contains two *Abies* species in its southern range; both go into dieback and are ecologically quite similar at this level. It is remarkable that when a third species, a birch, enters these subalpine fir forests, their dieback behavior disappears, and both *Abies* species increase in size and age. This is a clear example for the functional role of biodiversity in a particular life-form group.

The Galápagos and Hawaiian island rain forests are particularly vulnerable to canopy species displacement in their dieback phase. Alien tree species, imported from elsewhere, can easily invade such dieback stands, and thus become successional displacers (Jacobi et al. 1988). Therefore, dieback has special implications for conservation management.

While the shifting mosaics of the "Shimagare" forests appear to be in equilibrium with their environmental factors, this cannot be said so easily for the other three forest systems. If the area of these forests is continuously reduced by allocation to other land uses, and preservation efforts are directed only to the currently healthiest forest segments, the conservation value of such forests may be lost when they enter the dieback stage. At that time they can become displaced by biological invasions of alien trees, shrubs, or grasses. Parts of the *Scalesia* forest in the Galápagos and the *Metrosideros* forest in Hawaii are already threatened by extinction of this sort. Natural forest systems consisting of a mosaic of different age states can maintain themselves, but if they are reduced to only one or two age states, their survival is threatened. Healthy, mature stands are needed for supplying disseminules. Therefore, systems with large mosaics require large territories, if preservation of natural processes is a goal of nature conservation.

In this chapter, I presented only four examples of Pacific forest biomes exhibiting natural dieback. A relationship of low canopy species diversity and spatial instability and the reverse can only be suggested but has not yet been established. As a general and testable hypothesis I propose that patch or gap sizes, other than those produced through obvious single-factor type 1 disturbances, are related to canopy species diversity and stand structure. Comparative research of patch dynamics, including a landscape-level approach to different biomes with low and high species diversities, would elucidate this hypothesis. Biotic impoverishment and the resultant structural simplicity are suggested as important causes of larger area instability.

Acknowledgements. I thank my wife, Annette Mueller-Dombois, for word processing and helpful discussions and Donald Drake for reviewing the manuscript. Financial support from the National Science Foundation for Research Grants BSR-841678 and BSR-871678 is gratefully acknowledged.

References

AIBS (1987) Tree death: cause and consequence. Bioscience 37 (8), A special feature issue with eight articles

Arentz F (1983) *Nothofagus* dieback on Mt. Giluwe, Papua New Guinea. Pac Sci 37:453–458

Arentz F (1988) Stand-level dieback etiology and its consequences in the forests of Papua New Guinea. GeoJournal 17:209–215

Arndt U, Mueller-Dombois D (eds) (1988) Forests of the world, stand-level dieback and ecosystem processes – a global perspective. GeoJournal 17 (2), Special Symposium issue with 22 papers

Ash J (1988) *Nothofagus* (Fagaceae) forest on Mt. Giluwe, New Guinea. New Zealand J Bot 26:245–258

Atkinson IAE (1970) Successional trends in the coastal and lowland forest on Mauna Loa and Kilauea Volcanoes, Hawaii. Pac Sci 24:387–400

Aubreville A (1938) La forêt coloniale: les forêt de l'Afrique occidentale française. Ann Acad Sci Colon 9:1–245

Bormann FH, Likens GE (1981) Pattern and process in a forested ecosystem. Springer, Berlin Heidelberg New York

Brokaw NVL (1985) Treefalls, regrowth, and community structure. In: Pickett STA, White PS (eds) The ecology of natural disturbance and patch dynamics. Academic Press, Lond New York, pp 53–69

Christensen NL (1981) Fire regimes in southeastern ecosystems. In: Mooney HA, Bonniksen TM, Christensen NL, Lotan JE, Reiners WA (eds) Fire regimes and ecosystem properties. USDA Forest Service Gen Tech Rept WO-26:112–136

Davison EM (1988) The role of waterlogging and *Phytophthora cinnamomi* in the decline and death of *Eucalyptus marginata* in Western Australia. GeoJournal 17:239–244

Hamann O (1979) Dynamics of a stand of *Scalesia pedunculata* Hook fil., Santa Cruz Island, Galápagos. Bot J Linn Soc 78:67–84

Hartshorn GS (1978) Treefalls and tropical forest dynamics. In: Tomlinson PB, Zimmermann MH (eds) Tropical trees as living systems. Cambridge Univ Press, Lond New York, pp 617–638

Heinselman ML (1981) Fire intensity and frequency as factors in the distribution and structure of northern ecosystems. In: Mooney HA, Bonniksen TM, Christensen NL; Lotan JE; Reiners WR (eds) Fire regimes and ecosystem properties. USDA Forest Service Gen Tech Rept WO-26:7–57

Holloway JD (1977) The Lepidoptera of Norfolk Island. Junk, The Hague

Holt RA (1983) The Maui forest trouble: a literature review and proposal for research. Hawaii Bot Sci Pap 42, 67 p

Holt RA (1988) The Maui forest trouble: reassessment of an historic forest dieback. M Sci Thesis, Univ Hawaii, Honolulu

Itow S, Mueller-Dombois D (1988) Population structure, stand-level dieback and recovery of *Scalesia pedunculata* forests in the Galápagos Islands. Ecol Res 3:333–339

Jacobi JD, Gerrish G, Mueller-Dombois D (1983) 'Ohi'a dieback in Hawaii: vegetation changes in permanent plots. Pac Sci 37:327–337

Jacobi JD, Gerrish G, Mueller-Dombois D, Whiteaker L (1988) Stand-level dieback and *Metrosideros* regeneration in the montane rainforest of Hawaii. GeoJournal 17:193–200

Jane GT, Green TGA (1983) Episodic forest mortality in the Kaimai Ranges, North Island New Zealand. New Zealand J Bot 21:21–31

Jones EW (1945) The structure and reproduction of the virgin forests of the north temperate zones. New Phytol 44:130–148

Kastdalen A (1982) Changes in the biology of Santa Cruz Island between 1935 and 1965. Noticias de Galápagos 35:7–12

Kohyama T (1988) Etiology of "Shimagare" dieback and regeneration in subalpine *Abies* forests of Japan. GeoJournal 17:201–208

Lawesson JE (1988) Stand-level dieback and regeneration of forests in the Galápagos Islands. Vegetatio 77:87–93

Leibundgut H (1959) Über Zweck und Methoden der Struktur-und Zuwachsanalyse von Urwäldern. Schweiz Z Forstwes 110:111–124

Leibundgut H (1978) Über die Dynamik Europäischer Urwälder. Allg Forstz 24:686–690

Lieberman M, Lieberman D, Peralta R (1989) Forests are not just Swiss cheese: canopy stereogeometry of non-gaps in tropical forests. Ecology 70:550–552

Lyon HL (1909) The forest disease on Maui. Hawaiian Planter's Record 1:151–159

Lyon HL (1919) Some observations on the forest problems of Hawaii. Hawiian Planter's Record 21:289–300

Matson PM, Boone R (1984) Natural distrubance and nitrogen mineralization: wave-form dieback of mountain hemlock in the Oregon Cascades. Ecology 65:1511–1516

Matson PM, Waring RH (1984) Effects of nutrient and light limitation on mountain hemlock: susceptibility to laminated root rot. Ecology 65:1517–1524

Mayer H, Neumann M (1981) Struktureller und entwicklungsdynamischer Vergleich der Fichten-Tannen-Buchen-Urwälder Rothwald/Niederösterreich und Corkova Uvala/Kroatien. Forstwiss Centralbl 100:111–132

Mayer H, Neumann M, Sommer HG (1980) Bestandsaufbau und Verjüngungsdynamik unter dem Einfluss natürlicher Walddichten im kroatischen Urwaldreservat Corkova Uvala/Plitvicer Seen. Schweiz Z Forstwes 131:45–70

Mueller-Dombois D (1983) Canopy dieback and successional processes in Pacific forests. Pac Sci 37:317–325

Mueller-Dombois D (1985) 'Ohi'a dieback in Hawaii: 1984 synthesis and evaluation. Pac Sci 39:150–170

Mueller-Dombois D (1986) Perspectives for an etiology of stand-level dieback. Ann Rev Ecol Syst 17:221–243

Mueller-Dombois D (1987) Natural dieback in forests. BioScience 37:575–583

Mueller-Dombois D (1988) Canopy dieback and ecosystem processes in the Pacific area. In: Proc XIV Int Bot Congr Koeltz, Königstein/Taunus, pp 445–465

Mueller-Dombois D, Ellenberg H (1974) Aims and methods of vegetation ecology. Wiley New York

Mueller-Dombois D, McQueen DR (eds) (1983) Canopy dieback and dynamic processes in Pacific forests. Pac Sci 37 (4), Spec Symp issue with 17 papers

Mueller-Dombois D, Jacobi JD, Cooray RG, Balakrishnan N (1980) 'Ohi'a rain forest study: ecological investigations of the 'ohi'a dieback problem in Hawaii. College of Trop Agric and Human Resources. Hawaii Ag Expt Sta Miscell Public 183:64

Mueller-Dombois D, Bridges KW, Carson HL (eds) (1981) Island ecosystems: biological organization in selected Hawaiian communities. US/IBP Synthesis Series 15. Hutchinson-Ross, Woods Hole, Massachusetts

Ogden J (1988) Forest dynamics and stand-level dieback in New Zealand's *Nothofagus* forests. GeoJournal 17:225–230

Palzer C (1983) Crown symptoms of regrowth dieback. Pac Sci 37:465–470

Papp RP, Kliejunas JT, Smith RS Jr, Scharpf RF (1979) Association of *Plagithmysus bilineatus* (Coleoptera: Cerambycida) and *Phytophthora cinnamomi* with the decline of 'ohi'a lehua forests on the Island of Hawaii. Forest Sci 25:187–196

Petteys EQP, Burgan RE, Nelson RE (1975) Ohia forest decline: its spread and severity in Hawaii. USDA Forest Serv Res Paper PSW-105. Pac SW For & Range Expt Sta, Berkeley, CA, 11 pp

Pickett STA, White PS (eds) (1985) The ecology of natural disturbance and patch dynamics. Academic Press, Lond New York

Platt WJ, Strong DR (1989) Special feature: "gaps in forest ecology". Ecology 70:535

Porter JR (1973) The growth and phenology of *Metrosideros* in Hawaii. Hawaii IBP Tech Rep No 24, Univ Hawaii, Honolulu, 62 pp

Schowalter TD (1985) Adaptations of insects to disturbance. In: Pickett STA, White PS (eds) The ecology of natural disturbance and patch dynamics. Academic Press, London New York, pp 235–252

Selling OH (1948) Studies in Hawaiian pollen statistics. III. On the late Quarternary history of the Hawaiian vegetation. Bernice P Bishop Mus Spec Publ Honolulu 39, 154 pp, 27 plates

Skipworth JP (1983) Canopy dieback in a New Zealand mountain beech forest. Pac Sci 37:391–395

Sprugel DG (1976) Dynamic structure of wave-regenerated *Abies balsamea* forests in the northeastern United States. J Ecol 64:889–911

Veblen TT (1985) Stand dynamics in Chilean *Nothofagus* forests. In: Pickett STA, White PS (eds) The ecology of natural disturbance and patch dynamics. Academic Press, Lond New York, pp 35–57

Walker J, Thompson CH, Jehne W (1983) Soil weathering stage, vegetation succession, and canopy dieback. Pac Sci 37:471–481

Wardle JA, Allen RB (1983) Dieback in New Zealand *Nothofagus* forests. Pac Sci 37:297–404

Watt AS (1947) Pattern and process in the plant community. J Ecol 35:1–22

White TCR (1986) Weather, *Eucalyptus* dieback in New England, and a general hypothesis of the cause of dieback. Pac Sci 40:58–78

Whitmore TC (1989) Canopy gaps and the two major groups of forest trees. Ecology 70:536–538

Zuckrigl K, Eckhardt G, Nather I (1963) Standortskundliche und waldbauliche Untersuchungen in Urwaldresten der niederösterreichischen Kalkalpen. Mitt Forstl Bundes-Versuchsanst Maria-brunn, Vienna Austria

Mosaic Cycles in the Marine Benthos

K. Reise

1 Introduction

In community succession, temporal changes are caused by a combination of
autogenic and allogenic, deterministic and stochastic processes (Tansley 1920;
Pickett and McDonnell 1989). Where such a succession starts from virgin sites, a
progressive sere may be observed. Sooner or later, however, patches undergo
cyclic changes, drift out of phase and a mosaic of species or age classes becomes
established (Fig. 1; Cooper 1923; Watt 1947; Remmert 1985).

In these mosaic patterns, populations find refuges, may spread the risk of
survival, or may choose between coarse and fine-grained strategies (Birch 1971;
Den Boer 1968; Wiens 1976). This may lead to an overall stability of loosely
coupled elements, dispersed over several habitat patches. The patches undergo
cyclic or erratic changes, and consequently there is no need for stability which is
maintained by biotic feedback interactions.

In this respect, the mosaic-cycle concept (Remmert, this Vol.) constitutes a
problem shift. Local stability is no longer relevant. New topics are the temporal
and spatial scales of the cycling communities, and the processes driving the cycles
or affecting their shapes. In most cases, this perspective requires investigations of
a wider scale in space and time than is conventionally adopted.

As a consequence, this review on mosaic cycles in marine benthic communities relies to some extent on rather circumstantial and indirect evidence.
Cycles are expected where late successional species occur in patches of single age
classes without recruits. Periodic outbreaks of grazers, predators or diseases are
often associated with cyclic successions. Recurrent physical disasters (Harper
1977) may be essential to continue cyclic development. My discussion is restricted
to large species. These often modify or even generate habitat structure, and utilize
a disproportionately large share of resources (Brown and Maurer 1986).

2 Rocky Shores

Compared to terrestial plant communities, sessile rocky shore organisms are a
rootless assemblage. There is no soil with a susceptible water level and a
depletable pool of nutrients. Barnacles, mussels and kelp attach to the rocks but

Biologische Anstalt Helgoland, Wattenmeerstation Sylt, 2282 List, Germany

H. Remmert (Ed.)
Ecological Studies Vol. 85
© Springer-Verlag Berlin Heidelberg 1991

├───── Progressive sere ─────┤├──── Mosaic cycles ────┤

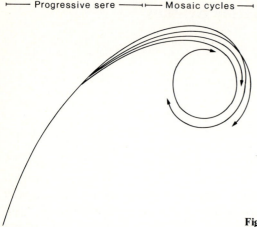

Fig. 1. Trajectory of community succession

mostly do not interact with this substrate. Food and nutrients are supplied by currents in variable amounts, and cannot be depleted in the long run. Similarities with terrestial plant communities include a potentially limiting primary space, competition for light in the case of algae, and predators or grazers may exert a strong influence on the structure of the community (Connell 1972; Dayton 1975; Paine 1974).

Since there are few or negligible modifications of the substrate, the progressive sere of succession (Fig. 1) will be unimportant, while cyclic changes are expected to predominate.

Connell (1975) summarized the processes driving cyclic changes in harsh, intermediate and benign rocky shore environments. Generally, there are two types of cycles, a short and a long one (Fig. 2). The short cycle simply represents repeated failures of colonizers in harsh environments, such as the upper intertidal zone. Young colonists regularly fell victim to desiccation, frost or other physical factors. In benign environments, such as the lower tidal zone, grazers and predators are abundant enough to consume all early settlers. Under intermediate conditions and on rare occasions in extreme environments, the 'bottleneck' of the short cycle is passed. The young colonizers achieve sizes at which they are invulnerable to most grazers or predators, and at which they can withstand desiccation to some extent.

Where several species managed to pass the bottleneck, usually a single dominant suppresses, displaces or excludes all other colonists in the patch, and a "monoculture" (Paine 1984) becomes established. The eventual destruction of such a monospecific patch may take various forms. The risk of disruption may be increased by the dominants themselves. Paine and Suchanek (1983) showed that larger tunicate individuals, like larger trees, are more susceptible to storm waves than smaller ones. In the absence of predation on small mussels, *Mytilus californianus* develops multilayered beds. These are disrupted more often by

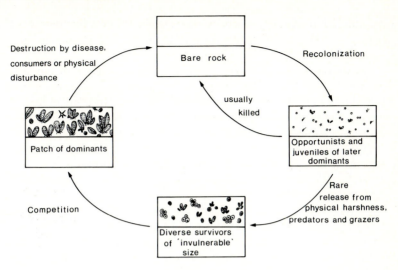

Fig. 2. Successional cycle of sessile rocky shore organisms with an ample supply of recruits from the sea (Modified after Connell 1975)

wave force than the structural monolayers which prevail in the presence of predatory snails and a small carnivorous starfish (Paine 1976).

The most frequently observed external process removing patches of dominants is clearance by storms. Menge (1976) estimated at a New England shore that up to 90% of the primary space present was cleared of sessile animals by storms during late winter and early spring. There are also a few predators which prefer the patches with the largest prey. Sea otters forage on mussels by removing entire clumps, discard small mussels and debris, while large mussels are broken open by pounding against a rock or another mussel balanced on their chest (VanBlaricom 1988). Similarly, humans tend to pick up the largest edible food organisms they can find on the shore (Moreno et al. 1984; Castilla and Bustamante 1989).

Although there is a general trend within patches towards monospecific spatial dominance with large and long-lived individuals, there is also a great deal of variability. Often cycles are not completed, or patches of a dominant are followed by young settlers of the same species, or settlement of larvae or spores is too scattered to make species interactions possible (Lewis 1977; Connell 1985; Roughgarden et al. 1985).

The apparent spatial kaleidoscope (Paine 1976) on most rocky shores, consisting of bare gaps, patches of young colonizers, single age classes or monocultures, may originate from various processes. Larvae show considerable patchiness during planktonic life in the tidal waters as well as when settling (de Wolf 1973). Rock topography and tidal level are other important causes of patchiness (Lewis 1964). Other common processes are gap formations by fluid-dynamic lift forces of breaking waves (Denny 1987) or by predators and grazers (Menge 1983), and by the dynamics of recovery. These processes are part of the cycle shown in Fig. 2.

Recolonization phases may contain several fugitive or pioneering species (Paine 1979, Sousa 1979). Turner (1983) studied algal succession on replicate plots and found that these soon ran out of phase. On some plots early colonists were still present, while on others they had been replaced by different species of later succession. Here, stochastic events generate a mosaic pattern. Finally, the historical sequence of events causes further patchiness. Gaps formed in different seasons tend to be colonized by different species (Osman 1977; Turner 1983). Episodic recruitment pulses in populations of long-lived individuals have lasting impacts on the subsequent development at a site.

Compared to terrestrial vegetation cycles, those on rocky shores seem to be faster (Paine 1980). In most cases patch longevity will match the life span of the spatial dominants. In algae this is between 2 to 4 years (Dayton 1975; Sousa 1979) while sessile invertebrates may live much longer, 10 to 20 years in barnacles (Connell and Sousa 1983) and up to 100 years in mussels (Suchanek 1981).

3 Kelp Forests

Along cool and temperate rocky coasts, large algae often form "submerged forests" from about the low tide line down to 20 or 30 m depth. These kelp forests tend to show mosaic patterns. There are gaps with barren rocks, multispecies patches composed of annual species and young of perannual species, and dense stands of a single, perannual canopy species (Dayton 1985). These patches constitute phases of a successional cycle (Fig. 3).

Herbivorous sea urchins play a key role. Usually they live in crevices between rocks, and feed on ephemeral algae or on shedded thalli of the large kelp. They are themselves prey to crabs, lobsters, fish or sea otters among others. Episodically their populations explode, either triggered by favourable conditions for the planktonic larvae (Foreman 1977) or by a release from benthic predation pressure (Mann 1977).

With increasing density, sea urchins move out of their refuges and form aggregations. They chew up all macroalgae they come across, and thus create gaps in the kelp forests. The aggregations then coalesce into coherent fronts and remove the entire vegetation.

In Nova Scotia kelp forests were overgrazed and destroyed along more than 500 km of coastline within 12 years by such sea urchin fronts (Mann 1982). Sea urchins are then able to persist on barren ground by browsing on epilithic microalgae and prevent the re-establishment of mature kelp stands (Chapman 1981). However, in the early 1980s these sea urchins suffered severe mortality by an epidemic disease. Their recovery was slow and spatially variable (Miller 1985; Scheibling 1986). Within 3 years of succession the kelp forests became re-established in those areas where urchins were decimated. Fishermen in the area remembered mass mortalities of sea urchins also from earlier decades of this

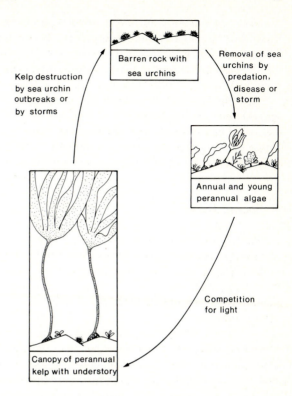

Fig. 3. Successional cycle of kelp
forests

century. Thus, there may be a repetitive seaweed-sea urchin cycle along the coast
of Nova Scotia with different phases in the various bays.

On the Pacific coast of North America, sea urchins are the preferred prey of
sea otters. In areas where these otters have been re-introduced, sea urchin density
declined and kelp forests developed (Duggins 1980; Estes and Harrold 1988). In
southeast Alaska, mosaic patterns within kelp forests are created by a large
starfish which preys on sea urchins (Duggins 1983). The starfish consumes only a
few individuals, but the rest exhibits a strong escape response. This creates
patches temporarily free of sea urchins. Here, a rapid algal succession takes place.
A new canopy becomes established because kelp attains a refuge as long as sea
urchins are restricted to crevices.

Next to sea urchins, storms play an important role in the dynamics of kelp
forests (Dayton and Tegner 1984; Ebeling et al. 1985). In southern California, a
severe storm removed the canopy of the giant kelp, *Macrocystis pyrifera*, but
spared the understory algae. Before the storm, sea urchins fed on detached blades
of the giant kelp. After the storm their preferred food had diminished, thus they
emerged from their shelters and started to consume all algae which had survived
the storm. The former kelp forest turned into barren ground with abundant sea
urchins on the rock surfaces.

After 3 years a second severe storm struck the site. This time the exposed sea urchins were removed, and the kelp forest became re-established within 2 years. The whole cycle was mediated by storms which had opposite effects, depending on the state of the community. Before the first storm struck the site, a kelp forest had persisted for about 12 years.

During such a phase, competitive interactions between and within plants species affect the structure (Dayton et al. 1984; Dean et al. 1989). An overlying canopy of giant kelp shades smaller species and its own recruits. At canopy gaps, faster growing individuals survived and depressed those below them. This occurred even within cohorts of juvenile *Macrocystis pyrifera*.

4 Biogenic Reefs

Several organisms generate elevated structures (bioherms) on the seabed. Photosynthetic microorganisms trap and bind sediment and may give rise to meter-sized stromatolite columns or domes (Awramik and Vanyo 1986). Many sedentary invertebrates accumulate and stabilize sediment, but the most conspicuous bioherms are the massive calcium carbonate accretions of the coral reef communities.

A precursor of biogenic reefs are multispecies clumps of epibenthic macrofauna observed on the bottom of the Adriatic Sea (Fedra 1977). A mobile brittle star is dominant but altogether 88 taxa were found in various combinations. By aggregating into clumps, suspension feeders may tap the swifter currents some centimeters above the bottom, and other animals find food and shelter between them. Those individuals in the center, however, receive less food and oxygen, and the clumps disintegrate once a certain size (500 to 1000 g) is exceeded. Thus, there is a mosaic of differently sized clumps which are in various phases of clump formation and clump disintegration.

A 12-year cycle is described by Gruet (1986) for reefs built by a single species, the polychaete *Sabellaria alveolata*. In the macrotidal Mont Saint-Michel Bay (France) the sabellarian colonies may start as epigrowth on an oyster shell, then spread laterally over the sandy flat and may finally attain a reef height of up to 1.2 m. Through upward growth, reefs create conditions for their own erosion by hydrodynamic forces (Fig. 4). Colonies are formed by aggregative larval settlement. There is no asexual budding and colonies are composed of several generations. The young ones tend to live along the sides of the constructions. The life span of individual worms is usually between 3 to 5 years.

For a related species, Barry (1989) showed that individuals respond to disastrous wave disturbance by releasing gametes into the water. Individuals thus maximize reproductive effort when faced with an increase in both the probability of death and the availability of new settlement space. This leads to a renewal of the reef by the establishment of several tiny colonies. Over a period of about 6 years these merge into a coherent reef and attain again more than 1 m in height. One of the sabellarian reef complexes in Mont Saint-Michel Bay exists at least since 1916.

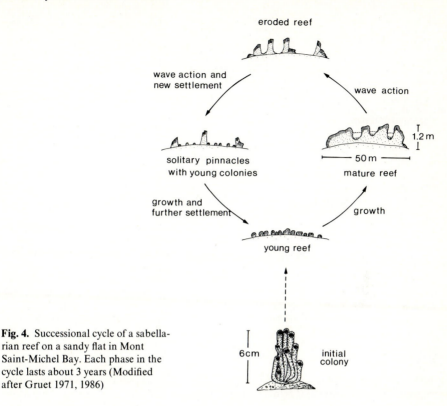

Fig. 4. Successional cycle of a sabellarian reef on a sandy flat in Mont Saint-Michel Bay. Each phase in the cycle lasts about 3 years (Modified after Gruet 1971, 1986)

A developmental cycle, which may be accomplished by a monospecific sabellarian reef in about 12 years, requires more than 1000 years in coral reefs. Mergner and Schuhmacher (1974) interpreted different shapes of a recent fringing reef in the Red Sea as different phases of a geological cycle (Fig. 5). Optimal growth conditions for hermatypic (= reef-building) corals are at the wave-exposed edge of the platform. This edge moves seaward by the addition of new structures.

With increasing depth, this process slows down and finally comes to an end because the corals at the upper reef front are unable to add sufficient material to the basement. The slope becomes steeper with overhanging parts. From now on erosive forces overcome the accretion rate. First, at the back reef corals are no longer cleaned by the surf, receive less food and have to endure the highest temperature fluctuations. These corals die and boring organisms make them ripe for erosive processes. These carve out a lagoon. On the seaward crest, erosion forms deep canyons and tunnels. Since the reef no longer grows, wave forces always attack the same parts. Eventually, the front of the reef breaks apart. A few sturdy pinnacles are left as remnants of the former crest.

As soon as the outer reef is obliterated and waves pass onto the shore, a new reef develops and starts again with seaward growth. Such perfect cyclic development, however, requires that the environmental conditions (climate, sea

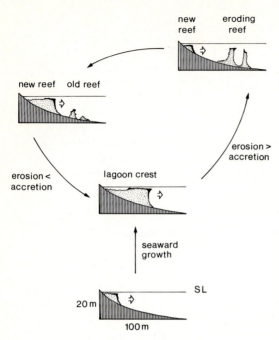

Fig. 5. Morphogenic cycle of fringing reefs in the Red Sea. Sea level (*SL*) is assumed to be constant. The cycle is driven by an imbalance between calcareous accretion by the living reef (*black*) and erosion of the dead reef rock (*dotted*) (Modified after Mergner and Schuhmacher 1974)

level, bedrock position, etc.) remain constant. This is rather unlikely over long periods of time, and most fringing reefs have a more irregular geological history (Adey 1978).

From these morphogenic changes which have formed the present coral reefs throughout the Holocene for about 9000 years B.P., we now turn to the ecological time scale which covers few to several generations of corals. To these changes the term succession is applied, and an overview is outlined in Fig. 6.

Most often, new substrate is first colonized by ephemeral algae (Benayahu and Loya 1977; Schuhmacher 1977). During the early months, also barnacles, oysters and other solitary animals settle. Upon the calcareous shells, coral planulae become attached, and within 1 year colonies grow about 1 cm in diameter.

Massive recruitment with carpets of young corals have not been reported. A paucity of settlers seems to be the rule, and these suffer heavy mortality until they attain a larger size which protects them against most grazers, many predators and dislocations (Connell 1973; Schuhmacher 1977; Sammarco 1980). Among corals, there are species which predictably colonize first (Loya 1976).

Coral assemblages with a great number of small colonies belonging to many different species, and with ample space left between them, are a common phenomenon. As an extreme example, Mergner and Schuhmacher (1981) recorded in the Red Sea at 10 m depth in a fore reef area of 5×5 m a total of 1274 colonies of stony corals (Scleractinia) and 749 colonies of soft corals (Alcyonaria). The average colony size was 52 cm². Scleractinia contributed 72 species and

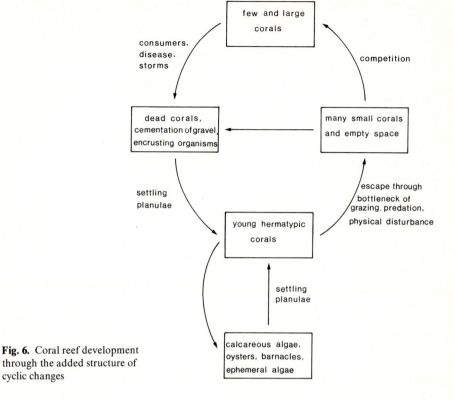

Fig. 6. Coral reef development through the added structure of cyclic changes

Alcyonaria 19. Empty space amounted to 56%. In such assemblages coral interactions play only a minor role in determining species composition and diversity (Bradbury and Young 1983). The supply with planula larvae and the "bottleneck" of juvenile survival seems to be decisive.

Particularly at the windward reef edge and the upper slope, corals may cover all available space. Here, coral growth is rapid, colonies are larger and diversity is low. Grigg (1983) even reports a case, where a single species (*Porites lobata*) occupied almost 100% of the reef surface. Growth forms with simple or compound plates or with long branches are capable of overtopping globose species, which in turn may defend themselves by extruding mesenterial filaments, thus digesting their competitors (Porter 1974; Lang 1973; Glynn 1976).

Physical disturbances are quite frequent on most coral reefs. Large colonies of the branching, foliose and plate-shaped type, particularly when already under attack from boring sponges or clams, are preferentially affected by the hydrodynamic forces (Chappell 1980). Generally, species that ordinarily win in competition suffer proportionately more from storm damage (Connell 1978). Coral plates also suffer under enhanced sedimentation. Globose and encrusting forms survive best prolonged low tide exposure.

Severe disturbances like cold spells, El Niño events, exceptional low water exposure and cyclones, which clear entire reef crests, are rare. Most disturbances have patchy effects and are moderate in intensity, extent and frequency. These prevent succession towards intense competition leading eventually to low diversity (Connell 1978; Grigg 1983).

The starfish *Acanthaster planci* is mostly a sporadic inhabitant of Indo-Pacific reefs. It consumes the living tissue of broken-off branches and kills only a few colonies. However, occasionally this echinoderm attacks coral reefs in vast numbers, leaving dead coral skeletons behind. Walbran et al. (1989) documented that such population outbreaks have occurred since the early formation of the Great Barrier Reef, about 8000 years B.P. They constitute an integral part of reef development.

Sea urchins keep algal growth on coral reefs within limits (Sammarco et al. 1974; Hay 1984). In the Carribean Sea, the sea urchin *Diadema antillarum* suffered wide-spread mortality from a water-borne pathogen in 1983 (Lessios et al. 1984). Following this disease, the cover with fleshy and filamentous algae increased (Ruyter van Steveninck and Bak 1986). This in turn was harmful to corals which then decreased in abundance.

The occasional outbursts and breakdowns of echinoderm populations affect some reefs but not others, and not all at the same time. This causes a large-scale mosaic of reefs which are in different phases of recovery. Estimates of recovery time after severe physical or biotic disturbances vary between 10 to 50 years (e.g. Stephenson et al. 1958; Grigg 1983).

Dead coral heads are more attractive to other reef organisms than living colonies. A 6-year-old live colony of *Pocillopora danae* was host to eight other species. By contrast, a dead colony of similar size and age contained 42 macroscopic species (Schuhmacher 1977). Dead corals are also a most suitable substrate for young corals. Wallace et al. (1986) found three to four juvenile corals (1 mm in size) per 100 cm² on a plate forming *Acropora*, whose tissues had been consumed by the starfish *Acanthaster planci* 2 months earlier. Of these juveniles 34% were still alive 1 year later and measured almost 1 cm in diameter. It seems likely that at least one of them will survive the gradual disintegration of the dead *Acropora* plate and will then occupy the vacant space.

A special feature of many shallow reefs are micro-atolls, which originate from a single globose colony (Fig. 7). Through upward growth, the top eventually suffers most from the stress of occasional low tide exposure. These large colonies die in their center while lateral growth is still proceeding. Bioerosion contributes to the formation of an enlarging lagoon area in the center. This now provides open space into which new corals can colonize. Mergner and Schuhmacher (1974) mapped a micro-atoll of *Platygyra lamellina* which was 7 m in diameter. This was a complex of three generations of nested micro-atolls, and four other coral species occurred also inside. Furthermore, three species of macroalgae and the encrusting *Lithothamnion* dwelled there. This is a perfect example of mosaic cycles on the scale of a single micro-atoll complex. Similar mosaic cycles occur at the scale between micro-atolls within a reef crest, and finally the concept also applies when reef crests within a region are considered (see above).

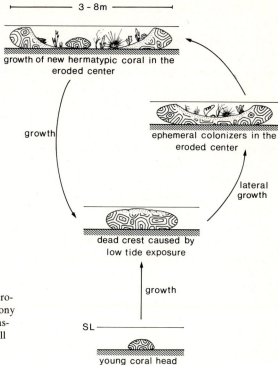

Fig. 7. Morphogenic cycle of a micro-atoll formed by a massive coral colony in shallow water. Sea level (*SL*) is assumed to be constant (After Connell 1978; Benayahu and Loya 1977; Mergner and Schuhmacher 1974)

5 Mud and Sand

About two-thirds of the earth is covered with marine sediments. Here, the most conspicuous organisms are not plants but mobile invertebrate animals. Nevertheless, we still find a corresponding pattern of a progressive succession which eventually turns into a mosaic of cycling patches.

A general scheme for succession from azoic sediment towards a diverse benthic assemblage has been proposed by Pearson and Rosenberg (1976, 1978), Rhoads et al. (1978), and Rhoads and Boyer (1982). Such a progressive sere may be observed spatially along an environmental gradient from high to low intensity or frequency of disturbance, e.g. a gradient from grossly polluted to normal sediment. Following a major disturbance (anoxia or a severe winter) the progressive sere may show up temporally.

The pioneering stage is usually dominated by small opportunistic worms (i.e. Spionidae, *Capitella*, Oligochaeta) feeding on the surface of the sediment or just below. They quickly establish dense populations, which are, however, ephemeral. Lack of intraspecific interference competition and the consequent depletion of food often cause sudden outbreaks followed by sudden breakdowns in these populations (Grassle and Grassle 1974; Chesney 1985).

Following the pioneer stage, gradually larger and deep-burrowing species establish themselves. Over time, they turn the sediment into a complex, three-dimensional fabric with tubes and burrows, fecal mounds and feeding funnels. They rework the sediment, flush overlying water in and out, cause sediment oxygenation, enhanced nutrient fluxes, maintain bacterial gardens and provide a variety of microhabitats for other, small species (Cadée 1976; Aller and Yingst 1978; Hylleberg 1975; Reise 1981).

This final phase of the progressive sere has been termed "equilibrium stage", implying indefinite persistence until a major disturbance event strikes the assemblage (Rhoads and Boyer 1982). However, the internal structure is very patchy. This is caused by two processes: biogenic habitat modification and patchy disturbances. Both may be viewed in the light of cyclic succession (Fig. 8).

Habitat patches come and go with the arrival and death of large tube-builders or burrowers. The anemone *Pachycerianthus multiplicatus* builds a tube, upon and within which 40 animal species can be found (O'Connor et al. 1977). Fecal mounds of the deposit-feeding holothurian *Molpadia oolitica* are preferentially colonized by a small, suspension-feeding polychaete (Rhoads and Young 1971). The different parts of lugworm burrows are differentially populated by a host of meiofaunal species (Reise and Ax 1979; Reise 1987). Aggregates of the snail *Umbonium vestiarium* form mud banks of 1 m² or more, which slowly extend over

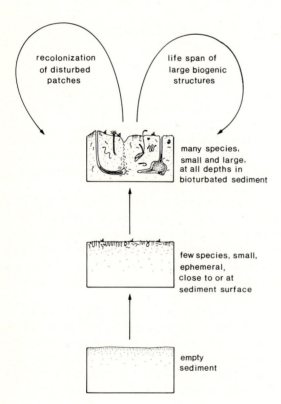

recolonization of disturbed patches

life span of large biogenic structures

many species, small and large, at all depths in bioturbated sediment

few species, small, ephemeral, close to or at sediment surface

empty sediment

Fig. 8. Progressive and cyclic changes of the macrofauna in marine sediments

sandy flats. In front, within and behind such snail banks, a different set of associated species was found (Hüttel 1986; pers. obser.). Structures of this type with various responses of an associated community are described from the intertidal zone down to the deep sea (Smith et al. 1986; Thistle and Eckman 1988; Grassle 1989).

These biogenic landscapes are bound to the fate of the individuals or aggregates which produce them. When these move, emigrate or die, this is a disaster to all the associates. Now these are also forced to move or die. A gap thus created may be inhabited first by various small species, until a new habitat-generating species arrives, modifying the local environment, and thus initiating the colonization of associates. Time of burrow establishment, age of burrow and characteristics of the local neighbourhood will all have an impact on the composition of the associated fauna. This is an obvious analogue to gap formation and gap recolonization in tropical rain forests (Connell 1978).

Patchy disturbances from biotic agents or physical forces are another source of mosaic patterns in marine sediments (Thistle 1981). Feeding rays form pits in the sediment, and it takes weeks until the benthic fauna at pit sites returns to the background composition (VanBlaricom 1982). Much larger pits are created by grey whales when feeding on amphipods at the bottom of the Bering Sea (Johnson and Nelson 1984). With such biotic and physical disturbances in mind, Johnson (1970, 1973) developed a model of the soft-bottom benthos viewed as a spatial and temporal mosaic. Each habitat consists of a mosaic of patches, each of which may be at a different stage of succession, either because of different ages of the patch or because the available recruitment species differed. Recolonization by mobile adults from neighbouring patches is also common (Thrush and Roper 1988; Frid 1989). Grassle and Sanders (1973) speculated that competitively inferior, fugitive species manage to persist in established communities by exploiting empty mosaic elements. This may occur but it is difficult to prove. In sediments each disturbance not only removes established species but also changes the texture, fabric and chemistry of the sediment. Thus, species colonizing first may simply have sediment requirements specific to freshly disturbed patches.

The above mosaic cycles are on the scale of a few centimeters to a few meters. Eagle (1975) described faunal dynamics on a spatio-temporal scale of kilometers and years (Fig. 9). In a shallow, wave-affected sandy region an assemblage dominated by the discretely mobile polychaete *Magelona papillicornis* persisted throughout 1970 to 1974 with only minor variations in seaward extension. In a deeper region, composed of sandy mud, there was a rapid turnover of single age-class populations of dominants. It is assumed that the deposit-feeders *Pectinaria koreni* and *Abra alba* inhibit further larval settlement once they are established. They also reduce sediment stability. Consequently, these populations either die naturally or an incidence of erosion takes them away. The area is then free for any settlers which happen to be in the area as larvae at the right time.

This study from Liverpool Bay suggests a bifurcation for species assemblages in marine sediments: type 1 persists (apparently) indefinitely and type 2 undergoes cyclic or erratic changes. The persistent type applies to various populations of habitat-generating worms, shrimp and bivalves. Lugworm

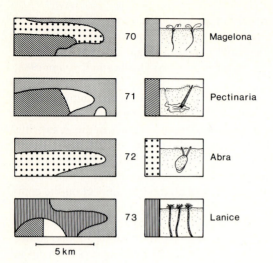

70   Magelona

71   Pectinaria

72   Abra

73   Lanice

5 km

Fig. 9. Variation in space (3.5 × 10 km) and time (October 1970 to October 1973) of faunal associations at about 10 m depth in Liverpool Bay. Associations are characterized by dominant species (*right*) (Modified after Eagle 1975)

populations in the harsh intertidal zone, for example, show a remarkable constancy and persistence (Reise 1985). Juveniles settle mainly at the edge or even outside of the adult population and migrate back when approaching adult size. Similar juvenile migrations are known from several benthic invertebrates, and also from fish or terrestrial animals. This space-for-time substitution allows for persistence of adult populations which exclude their young. In sessile invertebrates and plants such an adult dominance would lead inevitably to cycling. Sessile organisms are bound to temporal cycling, while mobile animals have the alternative to cycle spatially.

The changing type of assemblage, where dominance oscillates between two or more species, has been reviewed by Gray (1977) in view of neighbourhood stability or multiple stable points. Peterson (1984) and Sousa and Connell (1985) debated whether multiple stable states exist in nature. However, this question is rather academic in the face of the observed turnover of assemblages (Fig. 9). There is continuous change rather than phases of stability.

The causes of change usually involve a modification of sediment properties as a result of animal activities. In the case of the above-mentioned bivalve *Abra alba* and the polychaete *Pektinaria koreni* reduced sediment stability is caused by feeding movements and the accumulation of loose material in fecal mounds. Other large and mobile species act virtually as "bulldozers" and affect small, relatively immobile tube dwellers (Thayer 1983). Reduced sediment stability invites disaster. Strong waves or currents remove the entire assemblage.

Conversely, enhanced sediment stability by worm tubes or marine grasses with negative effects on freely burrowing species may also increase the chance of disaster over time. Accumulation of fecal material entails anoxia, and Den Hartog

(1970, 1985) observed pluriannual cycles in the intertidal patches of seagrass due to self-induced sedimentation which eventually interferes with photosynthesis.

Finally, changes may be caused by mobile predators such as crabs or fish which are particularly attracted to dense populations of their prey. Eagle (1975) explained the decline in the tube-dwelling polychaete *Lanice conchilega* (Fig. 9) as due to flatfish sucking the worms out of their tubes.

6 Scales of Mosaic Cycles in the Marine Benthos

Gray and Christie (1983) reviewed interannual cycles of temperature and salinity in the sea. They found periods of 3–4, 6–7, 10–11, 18–20 and 100 years. All could be related to astronomical periodicities. They also documented some correspondence between these periodicities and changes in benthic assemblages. In small amphipods and others, several generations are included within a single period. In some bivalves, single cohorts may last one or more decades. Long-term periodicities or rare disasters both terminate and give birth to such cohorts (Powell and Cummins 1985; Beukema 1989).

Thus, successional cycles will be affected by these pluriannual or decadal periodicities. Consequently, subsequent cycles may be similar but never completely identical. The trajectory of the species assemblage incorporates history, and thus affects its future course. The cycles shown in Fig. 1 will have to be substituted by a cluster of spirals.

In the sea, short-term variability of physical factors is damped by the very large heat capacity of the ocean (Steele 1985). This and the long-term exchange rates between deep and near-surface waters (in the order of centuries) lead to relatively large amplitude changes at long-term scales. Thus, the variance of many physical factors gradually increases as the time interval increases from days to decades. This environmental characteristic of the sea has been termed "red noise" in analogy to the red light spectrum.

Steele and Henderson (1984) modelled predator-prey populations in the presence of red noise. They showed that decades with low prey densities regulated by predation pressure, alternate abruptly with decades of high prey densities which are essentially resource limited. It follows that marine populations may undergo dramatic changes on the scale of decades to centuries without a corresponding dramatic change in the physical environment. Embedded in a multi-species community, changes may be less abrupt and less conspicuous. Up to now there is little empirical knowledge on such long-term cycles in benthic communities. However, a spatial mosaic structure is likewise to be expected. For example, long-term changes in the benthic amphipod *Pontoporeia affinis* in different parts of the Baltic Sea were not in concert (Gray and Christie 1983).

The evolutionary response of organisms to the consequences of red noise is to produce a large number of larvae, combined with an extensive potential for dispersal (Steele 1985). This allows a wide-scale scanning of living conditions, not only within a region but to some degree also between geographical provinces. In the southeastern Pacific, El Niño events cause major disturbances in the ecosystems along the west coast of South America. Sea-surface temperature increases and the current regime is altered. These episodes are 2 to 10 years apart.

The 1982–1983 El Niño was the strongest climatic anomaly recorded in this century (Cane 1983). It was accompanied by large reductions in plankton, pelagic fish, guano birds and seals, kelp forests and mussel beds were destroyed, but on the other hand, several warm-water species spread south over thousands of kilometers (Arntz 1986). This shows the enormous potential of some marine species to exploit new environments over large distances, within very short periods of time.

For the mosaic cycle concept, this implies that in the marine benthos the mosaic elements not only occur at scales of centimeters to kilometers, but may extend over latitudes. At such distances, corresponding phases of cycles will differ increasingly. The trajectory in Fig. 1 is fanning out.

With increasing scale, the forcing functions of cycles turn from organisms to physical processes. Small scale, rapid cycling is primarily driven by the organisms themselves. Large-scale, slow cycling is mainly driven by physical periodicities. The wide-scale responses of some organisms preclude the operation of fine-tuned, deterministic reactions towards each other and towards those which lack the potential of wide dispersal. This will produce irregularities in successive cycles.

On shore, the "red noise" gradually turns into "white noise". Here, the environmental variability increases towards the high tide line, and remains constant as the time interval increases from very short to very long periods, up to about 100 years (Steele 1985). Organisms and communities are expected to cope with high, short-term variability and in so doing, to minimize also the effects of longer-term variations. An example may be the persistent shallow sand assemblage compared to the alternation of dominants in the deeper, muddy sand area described in the preceding chapter (Fig. 9).

In a physically strongly damped environment, such as major parts of the deep sea, most mosaic cycles are probably driven by the organisms, and the corresponding patches are small. In the intertidal zone, adaptations to the high variability may result in similar, small-scale mosaic cycles within persistent assemblages. At intermediate depths, in the typical red noise environment, physical forcing is often essential for cycling, and the corresponding mosaic patches may be quite large.

7 Conclusions

Mosaic cycles are apparent at several scales in space and time from marine hard and soft bottoms, from biogenic reefs, and from the intertidal zone down to the deep sea. Successional changes in benthic macroalgae and invertebrates, both sessile and mobile, occur in patches which constitute different phases of cyclic development.

However, mosaic cycles are not universally present. Some shallow water assemblages, well adapted to cope with strong environmental variability, are often highly persistent. Often adult populations are replenished by adults, because their juveniles mature in outside nurseries before they fill the gap among the adults. Migration in space substitutes for turnovers in time. Also, these populations may show some degree of cyclicity at the scale of the ambits of its larger individuals or colonies. At a similar scale, mosaic cycles seem to occur in the deep sea. Here, organisms are adapted to a damped variability.

Mosaic cycles at intermediate scales are common where benthic organisms modify their habitat in a way which gradually increases the likelihood of being affected by rare physical disturbances such as heavy storms. This may occur by upward growth from the bottom of the sea, by accumulating mounds of silt or by destabilizing the sediment.

Many benthic species are capable of wide-range larval dispersal. This allows them to utilize habitat mosaics at very large spatial scales. It also allows them to shift to other geographical regions when confronted with an adverse phase of a long-term environmental periodicity.

In general, the cycling of small patches is usually driven biotically and tends to be short-term. Mosaic cycles at larger spatial scales are usually driven or mediated by physical processes and these tend to be long-term.

Compared to terrestrial vegetation, succession in the marine benthos is fast and in the order of 1 to 4 years. An exception is represented by coral communities which may need 50 years for full recovery from a major disaster (Grigg 1983). The period of cycles varies considerably, and is either limited by the longevity of dominants, which is usually well below 100 years, or by major periods of physical cycles on the scale of years and decades. Coral reefs may exhibit cyclicity on a geological time scale.

The application of the mosaic-cycle concept to the marine benthos is not new. It was formulated explicitly by Johnson (1970, 1973) for the macrofauna in sediments, and it is implicit in Connell's (1975, 1978) model for patches of dominants on rocky shores and for high coral diversity.

A somewhat similar concept is Sutherland's (1974) multiple stable points. A patch is viewed as stable until some disturbance pushes it into another stable state and so forth. In the mosaic cycle concept, the same patches are viewed as transient phases in a cyclic course of events. Patches are not in equilibrium. Change in desynchronized patches may bring apparent stability at a larger scale. Stability or instability is a question of the time scale chosen for observation.

Similarly, the notion of cycling trajectories for mosaic elements is an ideal which will rarely be met. With increasing spatial scale, mosaic patches will not be repetitive because of environmental gradients or regional differences. With increasing temporal scale, cycles are no longer repetitive because of long-term periodicities or trends in physical factors. Further, the history of past events is often reflected in the responses of species and communities to future events. Rather than perpetuating successional cycles, trajectories of community change will fan out into a cluster of spirals. Fractal symmetry decreases with increasing realism of the model.

The mosaic-cycle concept outlines a standard against which actual trajectories may be compared. It is not deterministic but provides a framework for the diversity of stochastic processes at the level of the mosaic elements. It emphasizes change rather than stability. It shifts attention from the progressive sere of succession to cyclic changes, and from early colonization to the decay or death phase of assemblages. The mosaic structure is no longer the domain of static analysis but its dynamic behaviour is now brought into focus.

In marine benthos monitoring, neither sampling a large area only once nor sampling isolated stations on a long-term basis is the proper strategy of investigation. Instead, repeated mapping of species assemblages within natural areas is required. The appropriate spatial and temporal scale of an investigation depends on the size and turnover rate of the mosaic elements, as well as the degree to which structures and processes are recurrent. An implication from the mosaic-cycle concept for the biological monitoring of marine benthos is to adopt a repeated mapping approach. This may help to differentiate between the natural cycling of mosaic elements and possible trends caused by pollution.

References

Adey WH (1978) Coral reef morphogenesis: a multidimensional model. Science 202:831–837
Aller RC, Yingst JY (1978) Biogeochemistry of tube-dwellings: a study of the sedentary polychaete *Amphitrite ornata* (Leidg). J Mar Res 36:201–254
Arntz WE (1986) The two faces of El Niño 1982–83. Meeresforsch 31:1–46
Awramik SM, Vanyo JP (1986) Heliotropism in modern stromatilites. Science 231:1279–1281
Barry JP (1989) Reproductive response of a marine annelid to winter storms: an analog to fire adaption in plants? Mar Ecol Prog Ser 54:99–107
Benayahu Y, Loya Y (1977) Space partitioning by stony corals, soft corals and benthic algae on the coral reefs of the northern Gulf of Eilat (Red Sea). Helgoländer Wiss Meeresunters 30:362–382
Beukema JJ (1989) Molluscan life spans and long-term cycles in benthic communities. Oecologia 80:570
Birch LC (1971) The role of environmental heterogeneity and genetical heterogeneity in determining distribution and abundance. In: den Boer PJ, Gradwell G (eds) Dynamics of populations. Cent Agric Publ Doc, Wageningen, The Netherlands, pp 109–128
Bradbury RH, Young PC (1983) Coral interactions and community structure: an analysis of spatial pattern. Mar Ecol Prog Ser 11:265–271
Brown JH, Maurer BA (1986) Body size, ecological dominance and Cope's role. Nature (Lond) 324:248–250
Cadée GC (1976) Sediment reworking of *Arenicola marina* on tidal flats in the Dutch Wadden Sea. Neth J Sea Res 10:440–460
Cane MA (1983) Oceanographic events during El Niño. Science 222:1189–1195
Castilla JC, Bustamante RH (1989) Human exclusion from rocky intertidal of Las Cruces, Central Chile: effects on *Durvillacea antarctica* (Phaeophyta, Durvilleales). Mar Ecol Prog Ser 50:203–214
Chapman ARO (1981) Stability of sea urchin dominated barren grounds following destructive grazing of kelp in St. Margaret's Bay, eastern Canada. Mar Biol 62:307–311
Chappell J (1980) Coral morphology, diversity and reef growth. Nature (Lond) 286:249–251
Chesney EJ (1985) Succession in soft-bottom benthic environments: are pioneering species really outcompeted? In: Gibbs PE (ed) Proc 19th Eur Mar Biol Symp. Cambridge Univ Press, Cambridge, pp 277–286
Connell JH (1972) Community interactions on marine rocky intertidal shores. Annu Rev Ecol Syst 3:169–192
Connell JH (1973) Population ecology of reef-building corals. In: Joanes OA, Endean R (eds) Biology and geology of coral reefs. Academic Press, Lond New York 2:205–246
Connell JH (1975) Some mechanisms producing structure in natural communities: a model and evidence from field experiments. In: Cody ML, Diamond JM (eds) Ecology and evolution of communities. Belknap Press, Cambridge, pp 460–490

Connell JH (1978) Diversity in tropical rain forests and coral reefs. Science 199:1302–1310

Connell JH (1985) The consequences of variation in initial settlement vs. post settlement mortality in rocky inter-tidal communities. J Exp Mar Biol Ecol 93:11–45

Connell JH, Sousa WP (1983) On the evidence needed to judge ecological stability or persistence. Am Nat 121:789–824

Cooper WS (1923) The recent ecological history of Glacier Bay, Alaska: II. The present vegetation cycle. Ecology 4:223–246

Dayton PK (1975) Experimental evaluation of ecological dominance in a rocky intertidal algal community. Ecol Monogr 45:137–159

Dayton PK (1985) Ecology of kelp communities. Annu Rev Ecol Syst 16:215–245

Dayton PK, Tegner MY (1984) Catastrophic storms, El Niño, and patch stability in the southern California kelp communit. Science 224:283–285

Dayton PK, Currie V, Gerrodette T, Keller B, Rosenthal R, Tresca DV (1984) Patch dynamics and stability of southern California kelp communities. Ecol Monogr 54:253–289

Dean TA, Thies K, Lagos SL (1989) Survival of juvenile giant kelp: the effects of demographic factors, competitors, and grazers. Ecology 70:483–495

Den Boer PJ (1968) Spreading of risk and the stabilization of animal numbers. Acta Biother 18:165–194

Den Hartog C (1970) The seagrasses of the world. Verh K Ned Akad Wet (Afd Natuurk 2 R) 59:1–275

Den Hartog C (1985) Factors effecting seagrass bed formation and breakdown (abstract). Estuaries 8:15A

Denny MW (1987) Lift as a mechanism of path initiation in mussel beds. J Exp Mar Biol Ecol 113:231–245

De Wolf P (1973) Ecological observations on the mechanisms of dispersal of barnacle larvae during planktonic life and settling. Neth J Sea Res 6:1–129

Duggins DO (1980) Kelp beds and sea otters: an experimental approach. Ecology 61:447–453

Duggins DO (1983) Starfish predation and the creation of mosaic patterns in a kelp-dominated community. Ecology 64:1610–1619

Eagle RA (1975) Natural fluctuations in a soft bottom benthic community. J Mar Biol Ass UK 55:865–878

Ebeling AW, Laur DR, Rowley RJ (1985) Severe storm disturbances and reversal of community structure in a southern California kelp forest. Mar Biol 84:287–294

Estes JA, Harrold C (1988) Sea otters, sea urchins, and kelp beds: some questions of scale. In: VanBlaricom GR, Estes JA (eds) The community ecology of sea otters. Springer, Berlin Heidelberg New York Tokyo, Ecol Stud 65:116–150

Fedra K (1977) Structural features of a north Adriatic benthic community. In: Keegan BF, O'Ceidigh PO, Boaden PJS (eds) Biology of benthic organisms. Pergamon Press, Oxford, pp 233–246

Foreman RE (1977) Benthic community modification and recovery following intensive grazing by *Strongylocentrotus droebachiensis*. Helgoländer Wiss Meeresunters 30:468–484

Frid CLJ (1989) The role of recolonization processes in benthic communities, with special reference to the interpretation of predator-induced effects. J Exp Mar Biol Ecol 126:163–171

Glynn PW (1976) Some physical and biological determinants of coral community structure in the eastern Pacific. Ecol Monogr 46:431–456

Grassle JF (1989) Species diversity in deep-sea communities. TREE 4:12–15

Grassle JF, Grassle JP (1974) Opportunistic life histories and genetic systems in marine benthic polychaetes. J Mar Res 32:253–284

Grassle JF, Sanders HL (1973) Life histories and the role of disturbance. Deep Sea Res 20:643–659

Gray JS (1977) The stability of benthic ecosystems. Helgoländer Wiss Meeresunters 30:427–444

Gray JS, Christie H (1983) Predicting long-term changes in marine benthic communities. Mar Ecol Prog Ser 13:87–94

Grigg RW (1983) Community structure, succession and development of coral reefs in Hawaii. Mar Ecol Prog Ser 11:1–14

Gruet Y (1971) Faune associée des "reciefs" édities par l'annélide *Sabellaria alveolata* (L.) en baje du Mont Saint-Michel: Banc des Hermelles. Mém Soc Sci Cherbourg 54:1–21

Gruet Y (1986) Spatio-temporal changes of sabellarian reefs built by the sedentary polychaete *Sabellaria alveolata* (Linne) PSZNI: Mar Ecol 7:303–319

80   K. Reise

Harper JL (1977) Population biology of plants. Academic Press, Lond New York
Hay ME (1984) Patterns of fish and urchin grazing on Carribean coral reefs: are previous results
typical? Ecology 64:446–454
Hüttel M (1986) Active aggregation and downshore migration in the trochid snail *Umbonium*
vestiarium (L.) on a tropical sand flat. Ophelia 26:221–232
Hylleberg J (1975) Selection feeding by *Abarenicola pacifica* with notes on *Abarenicola vagabunda* and
a concept of gardening in lugworms. Ophelia 14:113–137
Johnson KR, Nelson CH (1984) Side-scan sonar assessment of Gray Whale feeding in the Bering Sea.
Science 225:1150–1152
Johnson RG (1970) Variations in diversity within benthic marine communities. Am Nat 104:285–300
Johnson RG (1973) Conceptual models of benthic communities. In: Schopf TJM (ed) Models in
paleobiology. Freeman, Cooper and Co, San Francisco, pp 148–159
Lang LC (1973) Interspecific aggression by scleractinian corals. 2. Why the race is not only to the swift.
Bull Mar Sci 23:260–279
Lessios HA, Robertson DR, Cubit JD (1984) Spread of *Diadema* mass mortality throughout the
Carribean. Science 226:335–337
Lewis JR (1964) The ecology of rocky shores. English Univ Press, Lond
Lewis JR (1977) The role of physical and biological factors in the distribution and stability of rocky
shore communities. In: Keegan BF, O Ceidigh PO, Boaden PJS (eds) Biology of benthic
organisms. Pergamon Press, Oxford, pp 417–424
Loya Y (1976) The Red Sea coral *Stylophora pistillata* is an r-strategist. Nature (Lond) 259:478–480
Mann KH (1977) Destruction of kelp-beds by sea urchins: a cyclical phenomenon or irreversible
degradation? Helgoländer Wiss Meeresunters 30:455–467
Mann KH (1982) Kelp, sea urchins and predators: a review of strong interactions in rocky subtidal
systems of eastern Canada, 1970–1980. Neth J Sea Res 16:414–423
Menge BA (1976) Organization of the New England rocky intertidal community: role of predation,
competition, and environmental heterogeneity. Ecol Monogr 46:355–393
Menge BA (1983) Components of predation intensity in the low zone of the New England rocky
intertidal region. Oecologia 58:141–155
Mergner H, Schuhmacher H (1974) Morphologie, Ökologie und Zonierung von Korallenriffen bei
Aqaba (Golf von Aqaba, Rotes Meer). Helgoländer Wiss Meeresunters 26:238–358
Mergner H, Schuhmacher H (1981) Quantitative Analyse der Korallenbesiedlung eines Vorriffareals
bei Aqaba (Rotes Meer). Helgoländer Meeresunters 34:337–354
Miller RJ (1985) Succession in sea urchin and seaweed abundance in Nova Scotia, Canada. Mar Biol
84:275–286
Moreno CA, Sutherland JP, Jara FH (1984) Man as a predator in the intertidal zone of southern Chile.
Oikos 42:155–160
O'Connor B, Könnecker G, McGrath D, Keegan BF (1977) *Pachycerianthus multiplicatus* Carlgren
– biotope or biocoenesis? In: Keegan BF, Ceidigh PO, Boaden PJS (eds) Biology of benthic
organisms. Pergamon Press, Oxford, pp 475–482
Osman RW (1977) The establishment and development of a marine epifaunal community. Ecol
Monogr 47:37–63
Paine RT (1974) Intertidal community structure. Oecologia 15:93–120
Paine RT (1976) Size-limited predation: an observational and experimental approach with the
Mytilus – Pisaster interaction. Ecology 57:858–873
Paine RT (1979) Disaster, catastrophe, and local persistence of the sea palm *Postelsia palmaeformis*.
Science 205:685–687
Paine RT (1980) Food webs: Linkage, interaction strength and community infrastructure. J Anim
Ecol 49:667–685
Paine RT (1984) Ecological determinism in the competition for space. Ecology 65:1339–1348
Paine RT, Suchanek TH (1983) Convergence of ecological processes between independently evolved
competitive dominants: a tunicate-mussel comparison. Evolution 37:821–831
Pearson TH, Rosenberg R (1976) A comparative study of the effects on the marine environment of
wastes from the coastal industries in Scotland and Sweden. Ambio 5:77–79
Pearson TH, Rosenberg R (1978) Macrobenthic succession in relation to organic enrichment and
pollution of the marine environment. Oceanogr Mar Biol Annu Rev 16:229–311

Peterson CH (1984) Does a rigorous criterion for environmental identity preclude the existence of multiple stable points? Am Nat 124:127–133

Pickett STA, McDonnell MY (1989) Changing perspectives in community dynamics: a theory of successional forces. TREE 4:241–245

Porter JW (1974) Community structure of coral reefs on opposite sides of the Isthmus of Panama. Science 186:543–545

Powell EN, Cummins H (1985) Are molluscan maximum life spans determined by long-term cycles in benthic communities? Oecologia 67:177–182

Reise K (1981) High abundance of small zoobenthos around biogenic structures in tidal sediments of the Wadden Sea. Helgoländer Meeresunters 34:413–425

Reise K (1985) Tidal flat ecology. Springer, Berlin Heidelberg New York Tokyo

Reise K (1987) Spatial niches and long-term performance in meiobenthic Plathelminthes of an intertidal lugworm flat. Mar Ecol Prog Ser 38:1–11

Reise K, Ax P (1979) A meiofaunal "thiobios" limited to the anaerobic sulfide system of marine sand does not exist. Mar Biol 54:225–237

Remmert H (1985) Was geschieht im Klimax-Stadium? Naturwiss 72:505–512

Rhoads DC, Boyer LF (1982) The effects of marine benthos on physical properties of sediments. A successional perspective. In: McCall PL, Tevesz MJS (eds) Animal-sediment relations. Plenum Press, New York pp 3–52

Rhoads DC, Young DK (1971) Animal-sediment relations in Cape Cod Bay, Massachusetts. II. Reworking by *Molpadia oolitica* (Holothuroidea). Mar Biol 11:255–261

Rhoads DC, McCall PL, Yingst JY (1978) Disturbance and production on the estuarine sea floor. Am Sci 66:577–586

Roughgarden J, Iwasa Y, Baxter C (1985) Demographic theory for an open marine population with space-limited recruitment. Ecology 66:54–67

Ruyter van Steveninck ED de, Bak RPM (1986) Changes in abundance of coral-reef bottom components related to mass mortality of the sea urchin *Diadema antillarum*. Mar Ecol Prog Ser 34:87–94

Sammarco PW (1980) *Diadema* and its relationship to coral spat mortality: grazing, competition and biological disturbance. J Exp Mar Biol Ecol 45:245–272

Sammarco PW, Levinton JS, Ogden JC (1974) Grazing and control of coral reef community structure by *Diadema antillarum* Philippi (Echinodermata: Echinoidea): a preliminary study. J Mar Res 32:47–53

Scheibling R (1986) Increased macroalgal abundance following mass mortalities of sea urchins (*Strongylocentrotus droebachiensis*) along the Atlantic coast of Nova Scotia. Oecologia 68:186–198

Schuhmacher H (1977) Initial phases in reef development, studied at artificial reef types off Eilat, Red Sea. Helgoländer Wiss Meeresunters 30:400–411

Smith CR, Jumars PA, DeMaster DJ (1986) In situ studies of megafaunal mounds indicate rapid sediment turnover and community response at the deep-sea floor. Nature (Lond) 323:251–253

Sousa WP (1979) Experimental investigations of disturbance and ecological succession in a rocky intertidal algal community. Ecol Monogr 49:227–254

Sousa WP, Connell JH (1985) Further comments on the evidence for multiple stable points in natural communities. Am Nat 125:612–615

Steele JH (1985) A comparison of terrestrial and marine ecological systems. Nature (Lond) 313:355–358

Steele JH, Henderson EW (1984) Modeling long-term fluctuations in fish stocks. Science 224:985–987

Stephenson W, Endean R, Bennett I (1958) An ecological survey of the marine fauna of Low Isles, Queensland. Aust J Mar Freshw Res 9:261–318

Suchanek TH (1981) The role of disturbance in the evolution of life history strategies in the intertidal mussels *Mytilus edulis* and *Mytilus californianus*. Oecologia 50:143–152

Sutherland JP (1974) Multiple stable points in natural communities. Am Nat 108:859–873

Tansley AG (1920) The classification of vegetation and the concept of development. J Ecol 8:118–873

Thayer CW (1983) Sediment-mediated biological disturbance and the evolution of marine benthos. In: Tevesz MJS, McCall PL (eds) Biotic interactions in recent and fossil benthic communities. Plenum Press, New York Lond, pp 479–625

Thistle D (1981) Natural physical disturbances and communities of marine soft bottoms. Mar Ecol Prog Ser 6:223–228

Thistle D, Eckman JE (1988) Response of harpacticoid copepods to habitat structure at a deep-sea site. Hydrobiologia 167/168:143–149

Thrush SF, Roper DS (1988) Merits of macrofaunal colonization of intertidal mudflats for pollution monitoring: preliminary study. J Exp Mar Biol Ecol 116:219–233

Turner T (1983) Complexity of early and middle successional stages in a rocky intertidal surfgrass community. Oecologia 60:56–65

VanBlaricom GR (1982) Experimental analyses of structural regulation in a marine sand community exposed to oceanic swell. Ecol Monogr 52:283–305

VanBlaricom GR (1988) Effects of foraging by sea otters on mussel-dominated intertidal communities. In: VanBlaricom GR, Estes JA (eds) The community ecology of sea otters. Springer, Berlin Heidelberg New York Tokyo, Ecol Stud 65:48–91

Walbran PD, Henderson RA, Jull AJT, Head MJ (1989) Evidence from sediments of long-term *Acanthaster planci* predation on corals of the Great Barrier Reef. Science 245:847–850

Wallace CC, Watt A, Bul GD (1986) Recruitment of juvenile corals onto coral tables preyed upon by *Acanthaster planci*. Mar Ecol Prog Ser 32:299–306

Watt AS (1947) Pattern and process in the plant community. J Ecol 35:1–22

Wiens JA (1976) Population responses to patchy environments. Annu Rev Ecol Syst 7:81–120

Mosaic Distribution Patterns of Neotropical Forest Birds and Underlying Cyclic Disturbance Processes

J. HAFFER

1 Introduction

Due to the cyclic regeneration of forest vegetation (Aubreville 1938 cited in Richards 1952; Oldeman 1989) shifting small-scale mosaics of different habitat patches exist in tropical lowland forests, e.g., tree-fall gaps, overgrown forest edges, dense pioneer vegetation, and increasingly mature forest vegetation closing the gaps. Successional vegetation zones along shifting river courses and patches of different vegetation types on extensive floodplains provide habitat mosaics at a somewhat larger scale. These habitat mosaics, together with the structurally and floristically complex tropical forest vegetation, maintain the ecological heterogeneity necessary for the coexistence of the numerous species of neotropical forest birds. Climatic and paleoclimatic changes caused cyclic disturbance processes at a regional scale. Paleogeographic changes in the distribution of land areas and oceans (or inland seas) altered many distribution patterns on a continental scale.

In this review, I consider mosaic distribution patterns of neotropical forest birds at various hierarchical levels and discuss cyclic ecological disturbance processes which probably underlie the origin of these mosaic distribution patterns (Fig. 1). Ecologists study increasingly the distinctive nested combinations of disturbance, biotic response, and vegetation patterns at micro-, macro- and megascales in an attempt to understand landscape ecology (Delcourt et al. 1983; di Castri and Hadley 1988). Applying these concepts, Salo and Räsänen (1989) discussed the hierarchy of landscape patterns in western Amazonia.

2 Local Mosaic Distribution Patterns of Birds and Underlying Ecological Disturbance

In view of the complex habitat heterogeneity of tropical forests (Fig. 2) and the corresponding ecological adaptations of birds, highly varied mosaic distribution patterns of the numerous tropical species have developed at the local (micro-) level. The underlying ecological processes maintaining habitat heterogeneity and thus the local distribution patterns of birds include cyclic regeneration of the rain

Tommesweg 60, 4300 Essen 1, Germany

H. Remmert (Ed.)
Ecological Studies Vol. 85
© Springer-Verlag Berlin Heidelberg 1991

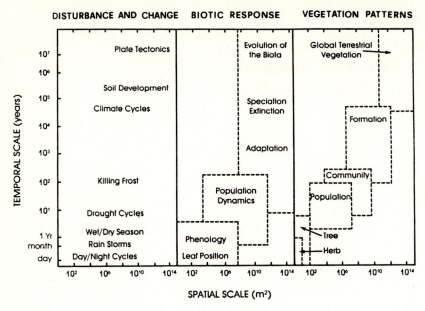

Fig. 1. Spatial and temporal interrelations of environmental disturbance and change, biotic response and vegetation patterns (After Delcourt et al. in Di Castri and Hadley 1988)

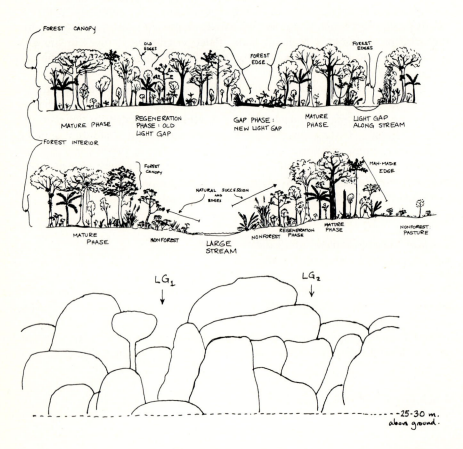

forest vegetation due to tree falls, local landslides, lateral river erosion, and forest fires. These site turnover processes represent free running regeneration cycles of the rain forest.

2.1 Local Mosaic Distribution Patterns

Rain forests on permanently dry ground (terra firme) consist of a shifting mosaic of habitat patches (eco-units, Oldeman 1989) which form a continuum from "mature phase" sites, with intact canopy, to "gap phase" (tree-fall) sites, with little or no canopy (Brokaw 1985). The sizes of tree-fall gaps usually range from 80 to 160 m² but occasional multiple tree-fall gaps are much larger (Fig. 3). The avifaunas of tree-fall gaps and the surrounding forest are quite distinct but their

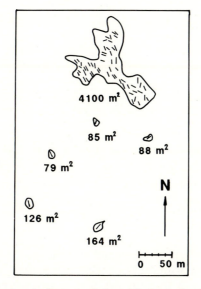
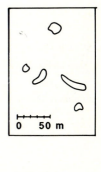

Fig. 3. Tree-fall gaps in rain forest consisting of "mature phase" forest with intact canopy and increasingly overgrown previous tree-fall gaps. *Left* A large multiple tree-fall gap, the closed canopy forest (*blank*), and five experimentally created single tree-fall gaps. *Dashes* in the gaps represent the gap-maker trees. Upper Rio Negro, near confluence with Caño Casiquiare, S Venezuela (modified after Uhl et al. 1988). *Right* Five naturally created tree-fall gaps in semideciduous, species-rich rain forest of central Panamá (modified after Schemske and Brokaw 1981)

Fig. 2. Rain forest in Costa Rica (after Stiles 1983). *Above* The forest mosaic: types of habitat. *Below* Configuration of the forest canopy at Finca La Selva, as seen from a platform 30 m above ground level. Note the presence of a sizable light gap of fairly recent formation (LG_1) and, in the foreground at *right*, an older gap now in the advanced regeneration phase (LG_2). Note also the amount of vertical "edge" habitat present in the canopy itself

compositions overlap considerably because gaps and mature forest phases are
end points on a local habitat gradient. In a study of the Costa Rican lowland forest
avifauna based on mistnet captures (Levey 1988a,b), 40% of the species with
adequate sample sizes (17 of 42 species) and 30% of the fruiting plant species (10
of 33) were found significantly more often in gaps than in intact forest (Table 1).
Only 5% of the bird species (2 of 42) and no plant species occurred more often in
intact forest. Many additional species were found more often in gaps than in
forest, and other species more often in forest than in gaps, but the differences
were statistically insignificant. The 17 gap specialists among birds consist of 10
frugivores, 4 nectarivores and 3 insectivores. Insectivorous birds are less sensitive
to ecological differences between gaps and forest than fruit-eating birds. Birds of
forest gaps also often inhabit old second growth forests and forest edges. However, these birds are not species of the canopy that follow the foliage/air interface
down into the gaps.

Table 1. Distribution of birds and understory fruiting plants in gaps and intact forest
sites in a Costa Rican rain forest (After Levey 1988a)[a]

	Birds		Fruit-eating birds		Fruiting plants	
	Gaps	Forest	Gaps	Forest	Gaps	Forest
No. species	77	60	30	19	83	58
No. individuals	997	637	490	267	427	206
No. habitat specialists[b]	17	2	10	1	10	0

[a] All gap/forest comparisons under each heading are significant at $P < 0.05$ except
number of species for birds and fruit-eating birds.
[b] A habitat specialist is defined as a species that was found significantly ($P < 0.05$)
more often in either gaps or intact forest.

Forest gaps are rich in fruiting plants and may represent important sources
of fruit during periods of fruit scarcity. More than twice as many plant individuals
were found fruiting in gaps than in the forest; ten species were fruiting
significantly more often in gaps than in forest and no plant species was found
fruiting more often in forest than in gaps. The local distribution patterns of
fruiting plants and frugivorous birds (Fig. 4) may reinforce each other in that the
birds disperse seeds into or around the periphery of the gaps more often than into
closed understory sites. Many plants fruit over longer periods in gaps than do
conspecifics in the forest, thus contributing to a larger fruit crop in forest gaps.
Large gaps maintain significantly higher densities of fruiting plants and fruit-
eating birds than small gaps. In Costa Rica, fruits and fruit-eating birds are most
abundant in the mid- to late rainy season (August-January) when north temperate and altitudinal migrants arrive (Fig. 5).

The avifauna of a less humid rain forest in central Panama comprised less gap
and forest specialists than the Costa Rican forest mentioned above. Schemske and

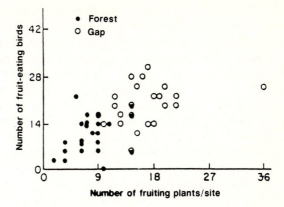

Fig. 4. Total numbers of fruit-eating birds and fruiting plants recorded along 46 net lanes that had complete data for 12 months. Humid lowland rain forest in Costa Rica (After Levey 1988a)

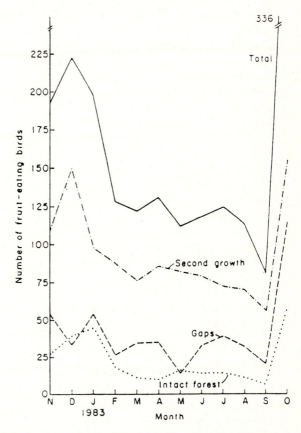

Fig. 5. Monthly captures of all fruit-eating birds in three habitats in humid lowland Costa Rica (After Levey 1988b)

Brokaw (1981) found only three gap specialists and two forest specialists among 31 species of adequate sample sizes. Fruit eaters in the gaps were rather un-common and fruit production was not very pronounced. However, gap and forest bird assemblages are quite distinct. The forest studied in Panama is partially deciduous and has a sparse canopy; thus, the distinction between gap and intact forest is less pronounced and may explain the differences to the Costa Rican forest avifauna.

Many neotropical forest bird species are widely distributed both geogra-phically and ecologically, occupying different habitats, whereas other species are ecological specialists and only occur in certain microhabitat patches (either over a large geographical region or in a rather restricted area). Such microhabitats and their inhabitants are, e.g., canopy vine tangles (*Cymbilaimus lineatus, Thamno-philus schistaceus, Cercomacra cinerascens*), *Heliconia* thickets in the forest (*Myrmeciza hyperythra*), vine-covered rotting logs from old tree-falls (*Microcerculus marginatus, Lioscelis thoracicus*) or tree-fall openings (*Thamnophilus aethiops, Hypocnemis cantator, Ornithion inerme, Syristes sibilator, Myrmothera campanisoma*), as mentioned by Terborgh et al. (1990) in their detailed quantitative study of the structure and organization of a western Amazonian forest bird community. This community consisted of 245 resident bird species which occupied all or part of a 97-ha plot of mature tropical floodplain forest in Amazonian Peru; 74 additional species visited the plot. Point diversities exceeded 160 species in some areas of the plot.

In regions dominated by active rivers, like western Amazonia, 20–40% of the area is periodically flooded either directly by the rivers or indirectly through a rise of the groundwater table (Fig. 6). The regional habitat mosaic in such river-dominated areas (Salo et al. 1986; Salo and Räsänen 1989) consists of (1) terra firme forest growing on old fluvial deposits (ca. 70% of the region); (2) complex forests on previous floodplains which are no longer reworked by rivers (ca. 15%); and (3) forest on the active floodplains of modern rivers (ca. 15%). The surface of young floodplains consists of a fine mosaic of sediment bars, flats, or oxbow lakes in varying stages of infilling and a complex mosaic of successional plant com-munities of riverine forest. The plant successional zones along the river courses in eastern Peru are sharply defined (Fig. 7; Terborgh 1985; Salo et al. 1986) and lead to conspicuous habitat selection of birds (Table 2):

1. Beaches and sandbars are mainly available outside the flood season.

2. Behind the open beach *Tessaria* trees grow in a zone 10 to 30 m wide; they mature in 3–4 years and reach a height of 8–10 m. Cane (*Gynerium*) forms dense stands locally.

3. River edge forest is dominated by tree genera such as *Cecropia, Ochroma,* and *Erythrina,* forming an open canopy of 14–18 m; *Heliconia* dominates the undergrowth.

4. The *Ficus-Cedrela* (várzea) forest forms a closed canopy and reaches a height of 40 m and an age of 100–200 years. *Heliconia* and palms dominate the undergrowth. This forest covers large areas.

5. Late successional forest is a mosaic forest that is seasonally inundated by rainfall or a rain-swollen stream, but not by a river.

6. High-ground forest grows on deep alluvium in irregular, disjunct patches within the floodplain where inundations are infrequent.

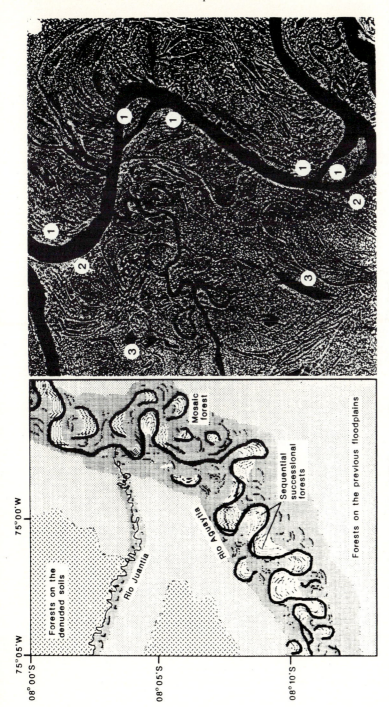

Fig. 6. River-dominated landscape in Amazonian Peru (after Salo et al. 1986). *Left* Simplified distribution map of forest types along Rio Aguaytia and Rio Juantia. The sequential successional forest is created on both sides of the meandering rivers. As the meandering proceeds, the sequential successional forest is cut from different angles and a mosaic forest structure is created.. During the annual flood cycle, sedimentation modifies the sequential meander scroll topography toward the flat topography of the former flood plain area. The previous floodplains are generally at a topographically higher level than the present floodplains. The denuded forest areas (*upper left corner*) representing the surface erosion relief pattern (convex-concave morphology) are laterally eroded by the Rio Juantia. The transitional forest (*lower right corner*) is developing towards the denuded pattern. *Right* Side-looking radar (SLAR) image of a floodplain mosaic. *1* Sites of intense primary succession at meander points leading to sequential successional forest. *2* Mosaic and transitional forest subject to lateral erosion at the outer curve of the meanders. *3* Isolated oxbow lakes

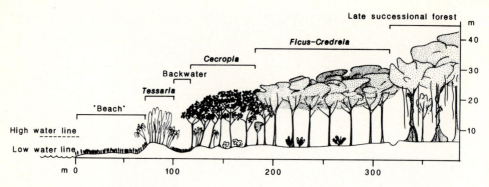

Fig. 7. Sequential succesional forest next to the Rio Manú at the Cocha Cashú biological station, eastern Perú (After Salo et al. 1986)

Birds of the late successional (mosaic) forest and the várzea forest (*Ficus-Cedrela* zone) are mainly species common to the high-ground forest; only a few species are restricted to these two successional stages (6 and 9%, respectively). The most distinctive communities with the highest proportion of exclusive species inhabit the ends of the successional gradient (Table 2): high-ground forest (40%) on one side, and the *Cecropia* (18%) to the *Tessaria* zone (62%) on the other extreme. Nectarivores and frugivores are absent from the *Tessaria* zone where no nectar or fruit are produced and where sallying-hovering insectivores are common. Habitats that are subject to seasonal inundation are deficient in terrestrial insectivorous birds (which are therefore conspicuous only in high-ground forest). Mixed flocks of omnivorous birds avoid the *Ficus-Cedrela* forest where none of the three canopy trees produce fruit that is eaten by small birds.

Furthermore, 102 bird species (15% of the land birds) are restricted to river-created habitats in Amazonia (Remsen and Parker 1983; Rosenberg 1990) of which 64 species occur in these habitats exclusively. The rest of the species also occupy secondary vegetation in Amazonia or also beyond in the Neotropical

Table 2. Total number of resident bird species, number of habitat exclusive species, and percentage of exclusive species in stages of riparian succession at Cocha Cashu, Rio Manu, eastern Peru (After Terborgh 1985)

	Successional stage				
	High-ground forest	Late successional forest	*Ficus-Cedrela* forest	*Gynerium-Cecropia* zone	*Tessaria*
Total number species	236	113	127	49	21
Number of exclusive species	92	7	12	9	13
Percentage of exclusive species	39	6	9	18	62

Region. Representatives of several species pairs replace each other in the riparian vegetation and terra firme forest, e.g. the nunbirds *Monasa nigrirostris/M. morphoeus,* the squirrel-cuckoos *Piaya cayana/P. melanogaster,* the antbirds *Myrmoborus lugubris/M. myotherinus,* the manakins *Schiffornis major/S. turdinus* (Snethlage 1913, p. 497, p. 538; Remsen and Parker 1983; Terborgh 1985, p. 332). Sixteen species restricted to river-created habitats represent monotypic genera indicating their systematically isolated position (e.g., *Anhima, Opisthocomus, Graydidasculus, Chelidoptera, Gymnoderus, Nasica, Sclateria, Myrmochanes, Donacobius, Ocyalus*). On the other hand, the geographical variation of the species found in river-created habitats is rather inconspicuous due to high mobility and gene flow within the populations inhabiting the vegetation zones along the river courses.

2.2 Cyclic Ecological Disturbance Processes

Ecological processes maintaining the habitat heterogeneity and local mosaic distribution patterns of neotropical forest birds include the cyclic regeneration of the forest due to gap-phase dynamics and fluvial dynamics (Table 3). Gaps in the forest leading to forest regeneration may be caused, e.g., by a single falling tree or several falling trees due to senescence, parasites, or illness (Richards 1952; Brokaw 1985; Whitmore 1989) or by local landslides in hilly regions following an earthquake (Fig. 8; Garwood et al. 1979). Three to 5% of terra firme forest is estimated to be in early successional stages following tree falls in gaps (Hartshorne 1980).

Multiple tree falls due, e.g., to strong wind, can affect a large portion of a forest. Turnover rates of lowland rain forest in Costa Rica due to gap-phase dynamics at different localities range from 80 to 140 years (Hartshorne 1978), indicating a very rapid rain forest regeneration.

Fluvial dynamics affects a fairly large portion (20–40%) of the landscape in western Amazonia and also leads to rapid forest turnover. The agents of river dynamics leading to cyclic forest regeneration (Fig. 9) are channel migration, channel diversion, and floodplain diversion (Salo and Kalliola 1989; Salo and Räsänen 1989). The time, during which a cycle is completed and the same site within a floodplain is again eroded by shifting river courses, may be only a few 100 years. This is due to the rapid lateral erosion of meandering rivers in eastern Peru (instances of 20 and even 250 m of riverbank erosion in 1 year have been observed). Turnover of forest sites due to river dynamics outside active floodplains may be in the order of 1000 or 2000 years.

In humid areas with a strongly seasonal climate the forest may become very dry toward the end of an abnormally dry season. Large patches of forest or certain plant species may die or dry out to such an extent that they burn when lightning ignites the forest (or a coal seam, oil shale, etc.). Fire may be considered as an intermediate disturbance for certain seasonal rain forests where fire occurs repeatedly, infrequently, and at low intensity (Sanford et al. 1985; Goldammer and Seibert 1989). When charcoal is found in association with ceramic artifacts,

Table 3. Hierarchical disturbance processes in the neotropical lowlands

Level of disturbance	Cause	Effect
I. Ecological disturbance		
1. Gap-phase dynamics	Senescence of trees, landslides, local denudation	Secondary succession, primary succession, denuded surfaces
2. Fluvial dynamics		
a. Channel migration	Lateral river erosion	Primary succession on point bar deposits
b. Channel diversion	Sudden lateral shift of river channel within floodplain (avulsion)	Primary and secondary succession on various types of floodplain deposits
c. Floodplain diversion	Aggradation or tectonic movements lead to sudden lateral shift of the river together with its floodplain	Primary and secondary succession on various types of floodplain deposits
3. Climatic dynamics	Extreme dry seasons lead to regional death of large patches of forest in some years; humid climatic extremes cause flooding of extensive areas	Fires burn forests in dry years (lightning may ignite forest, coal seams, or oil shales); strong winds cause multiple tree falls and flooding leads to the death of local forests
II. Historical disturbance		
4. Paleoclimatic dynamics	Regional climatic cycles (humid-warm, dry-cool; Tertiary-Quaternary)	Paleoclimatic changes in the composition of forests and in the distribution of forest and nonforest vegetation; large-scale flooding of lowlands
5. Tectonic dynamics	Regional uplift and subsidence due to epeirogenic and/or orogenic movements (Tertiary-Quaternary)	Paleogeographic changes in the distribution of land areas and flooded areas

Fig. 8. Hilly coastal area of SE Panama affected by an earth-quake and originally covered by undisturbed rain forest. The diagram illustrates a portion of an aerial photograph taken 3 days after an earthquake occurred showing the patchiness of landslides within the highly damaged central area. *Black areas* are landslides, *dashed lines* represent clouds on the photograph and *A* is an approximately 100-ha denuded area. The irregular lower edge is the Pacific coast line. The *top* is north; *bar* 1 km (After Garwood et al. 1979)

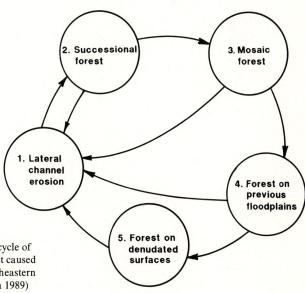

Fig. 9. Fluvial regeneration cycle of the western Amazonian forest caused by channel migration in southeastern Perú (After Salo and Kalliola 1989)

its presence is usually attributed to human occupation. However, charcoal also
occurs frequently in the soils of primary rain forest that has never been burned by
human inhabitants. The radiocarbon age of the charcoal under various forest
types in the San Carlos region of southern Venezuela ranged from 250 ± 50 years
B.P. to 6260 ± 110 years B.P. (Sanford et al. 1985; Saldarriaga and West 1986) and
under an East Kalimantan (Borneo) forest from 350 to 17510 years B.P.
(Goldammer and Seibert 1989).

3 Regional Mosaic Distribution Patterns of Forest Birds and Underlying Historical Disturbance Processes

3.1 Regional Mosaic Distribution Patterns

Many groups of closely allied neotropical forest bird species form large-scale
mosaic distribution patterns composed of neatly interlocking patches, which are
formed by the ranges of the component, representative species (Figs. 10–12).
Many of the representatives are biological species because of the absence or
near-absence of hybridization in the populations along the contact zones. The
representatives exclude one another geographically (or presumably do so) in
regionally rather uniform vegetation zones because the individuals of these
species are probably still strong ecological competitors. The representative
species in geographical contact are designated as parapatric (in contrast to
sympatric species, which inhabit the same region, and allopatric species whose
ranges are geographically separated).

The ecological inhomogeneities and obstacles present within the distribution
areas of each of the representative species, e.g., varying vegetation patterns, river
courses, etc., are evidently insufficient barriers to prevent the dispersion of these
species. These species are capable of either crossing these obstacles directly or
circumventing ecologically unfavorable areas, e.g. by following the banks of
broad rivers and their floodplains until reaching the opposite side. Forested
islands facilitate crossing of the river, and in the headwater region the rivers are
narrow and cease to be barriers. The lack of a barrier effect of many Amazonian
rivers may also reflect the fact that many rivers, except the widest ones, can change
course quickly, especially in flood season. In this way, even the most sedentary
birds and other animals of the forest interior are passively transferred across the
rivers and, thereby, continually reintroduce genes from one bank to the other.

However, the advance of representative species is stopped where they meet
a closely related ally that has advanced from a different direction. On the other
hand, the broad, lower portions of several Amazonian rivers in conjunction with
their wide floodplains are indeed effective barriers for birds inhabiting the forest
interior, and many species are unable to cross or circumvent these rivers, as
several previous authors have emphasized (e.g., Caparella 1988). This is particularly obvious in certain widely distributed species whose ranges are delimited
by a broad river and who have no close relatives occupying the opposite bank.
Thus, the lower Amazon River delimits, in the north, the ranges of the ground

cuckoo *Neomorphus geoffroyi* and the antbird *Myrmoborus myotherinus*. In these and other similar cases no representative species inhabit the northern bank of the lower Amazon.

An assemblage of allopatric and/or parapatric species is combined in a superspecies if the representatives are thought to be derived directly from conspecific ancestors (Mayr 1963; Amadon 1966; Haffer 1986). In zoogeographical analyses the superspecies and their ranges are more instructive than the individual distribution areas of their component species. Superspecies and systematically isolated (independent) species not belonging to a superspecies have been designated as "zoogeographical species" by Mayr and Short (1970; see also Bock and Farrand 1980). Consideration of superspecies puts the number of taxonomic species into a proper zoogeographical perspective. Thus, Haffer (1974) listed only 14 zoogeographical species in the toucans (Ramphastidae, from a total of 33 biological species) and 8 zoogeographical species in the jacamars (Galbulidae) with a total of 17 biological species. Zoogeographical species characterize the basic ecological units of a fauna and therefore have been designated "ecological species" by Bock and Farrand (1980). In this sense, parapatric species of a superspecies are still "conspecific" from an ecological point of view but, of course, they are distinct biospecies because they are genetically and reproductively isolated.

A few examples of parapatric birds are given below to illustrate the conspicuous regional mosaic distribution patterns of superspecies. Many additional examples from the Neotropical Region have been discussed in previous publications by Delacour and Amadon (1973) on curassows, Snow (1982) on cotingas, Paynter (1972, 1978) on finches of the genus *Atlapetes*, Willis (1969) on selected antbirds and Haffer (1970, 1974, 1977, 1987b) on ground cuckoos, jacamars, toucans, antbirds, cotingas, manakins, tanagers, and others.

The plumage of the *Ramphastos* toucans (Fig. 10) is mainly black, although some forms are tinged with maroon on the hindneck and back; the upper tail coverts are red, yellow, or white. Other differences between species and subspecies refer to the color of the throat and breast, the iris, and the bill. Two distinct and widely sympatric assemblages of parapatric species form mosaic distribution patterns over almost the entire Neotropical Region. These assemblages are composed of (1) medium-sized channel- or keel-billed species which have croaking vocalizations and (2) large, smooth-billed species with yelping calls. The latter are restricted to the canopy level, whereas the somewhat smaller channel-keel-billed toucans live mostly at somewhat lower levels in the forest. They are also more insectivorous than their larger relatives and are seen occasionally at ant raids near the forest floor or even on the ground (Willis 1983). Some of the geographical representatives of each group hybridize along their zones of contact, whereas others have reached the level of biospecies replacing each other parapatrically. For further details, see Haffer (1974).

The small *Selenidera* toucans (Fig. 11) form a similar mosaic distribution pattern in Amazonia. They inhabit dense, dark rain forest and are characterized by the black or chestnut underparts in males and females, respectively, and the yellow ear tufts. The dark-plumaged nunbirds (*Monasa*, Fig. 12) form a group of four closely related species of intermediate size. Two of these (*M. nigrifrons, M. morphoeus*) are sympatric in western and southeastern Amazonia where, however, they replace each other ecologically: *M. morphoeus* is a bird of the interior terra firme forest at midlevel to subcanopy heights (only occasionally occurring along forest borders; not along river borders). On the other hand, *M. nigrifrons* lives along forested borders of rivers, varzea, and second growth woodland (including woodland along the northern bank of the lower Amazon River!). Both species occur in small family groups. *M. morphoeus* und *M. nigrifrons* are replaced in NE Amazonia and in the Guianas by the black

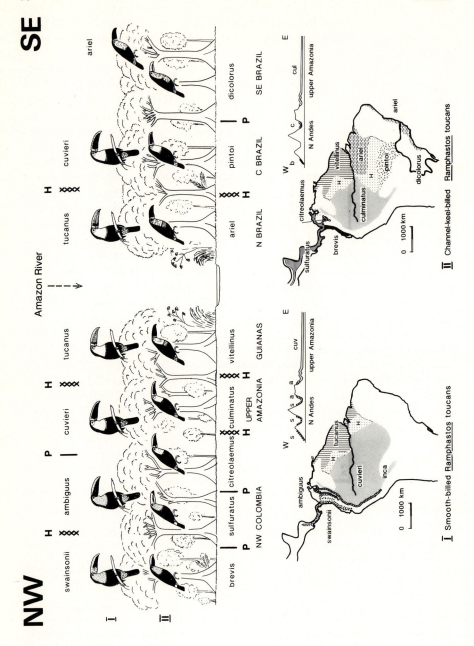

Fig. 11. Distribution of the lowland toucanets, *Selenidera maculirostris* superspecies (After Haffer 1974)

Fig. 10. Ecological occurrence and distribution of the *Ramphastos* toucans. **I** Smooth-billed "yelping" toucans (*upper figures* and *left-hand map*); **II** channel-keel-billed "grunting" toucans (*lower figures* and *right-hand map*). Plumage color of birds: *solid* black; *blank* white; *stippled* yellow; *dashed* red or orange. Relations of taxa along contact zones: *P* parapatry; *H* hybridization (hybrid zone). Abbreviations along schematic profiles across northern Andes (next to maps): *s swainsonii*; *a ambiguus*; *cuv cuvieri*; *b brevis*; *c citreolaemus*; *cul culminatus. Ramphastos toco*, a large species with a keeled bill which inhabits gallery forests in nonforest regions is not shown

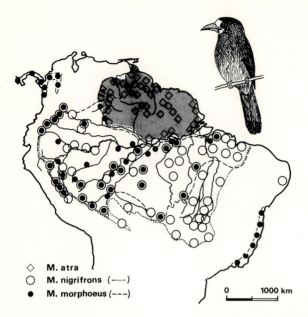

Fig. 12. Distribution of three species of nunbirds (*Monasa*, Bucconidae). See text for further explanations. The bird sketched is *M. morphoeus*

species *Monasa atra* with white wing marks, which lives in the rain forest and gallery forest as well as in second growth woodland, thus apparently preventing the range expansion of *M. morphoeus* and *M. nigrifrons* into NE Amazonia due to ecological competition. A fourth species, *M. flavifrons*, has a yellow bill (in contrast to the others with red bills) and inhabits forest borders and second growth woodland in the upper Amazonian foothill zone of the Andes.

Other closely allied sympatric bird species which, like *M. morphoeus*/*M. nigrifrons,* replace each other ecologically in Amazonia and inhabit the rain forest interior and vegetation zones along river borders, respectively, have been listed previously (Sect. 2.1).

3.2 Cyclic Historical Disturbance Processes

The representatives of a species assemblage forming a regional mosaic distribution pattern have been differentiated as subspecies and/or species during extended periods of geological time. The historical processes leading to this evolutionary differentiation contrast with the short-term ecological processes of gap-phase dynamics and river dynamics discussed above. However, the nature of these historical processes leading to the separation and differentiation of animal populations still remains unknown. Among the likely processes which have been proposed (Haffer 1969, 1974, 1982) are (1) paleogeographical changes in the distribution of land areas and regions covered by the sea during the Tertiary and Quaternary periods due to crustal movements of the earth or by eustatic sea level fluctuations and (2) climatic-vegetational changes during alternating humid and arid climatic phases of the Tertiary and Quaternary (Table 3). Both these processes are geologically constrained and essentially cyclic in nature. Many portions of the Neotropical Region were uplifted during geological history, thus becoming dry land, being worn down by erosion, and eventually again being subjected to subsidence. Similarly, many portions of the Neotropical Region have been affected by cyclic vegetation changes during the course of the history of the earth.

During climatic-vegetational cycles of the Cenozoic, the humid rain forest in a particular region was replaced several times by open palm forest and liana forest and possibly even savanna woodland before the vegetation cycle returned via open forest to humid rain forest (Fig. 13). Certain areas of the Neotropics have probably been affected more intensively by these vegetational changes than others where the rain forest was left more or less unchanged during a particular dry phase ("forest refuges"; see Fig. 14). Detailed geoscientific evidence from many regions in tropical America indicates that extensive climatic-vegetational shifts have taken place during the geological past (reviews by Tricart 1985; Bigarella and Ferreira 1985; Brown 1987; Haffer 1987a; Schubert 1988), indirectly supporting the notion of the refuge theory. Direct evidence that particular areas have supported a forest or nonforest refuge during one or more of the many climatic-vegetational reversals of the Tertiary and the Quaternary periods, however, is not yet available. Generally, it is probably too simplistic to assume

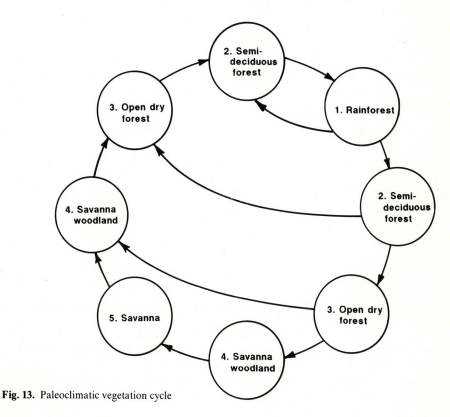

Fig. 13. Paleoclimatic vegetation cycle

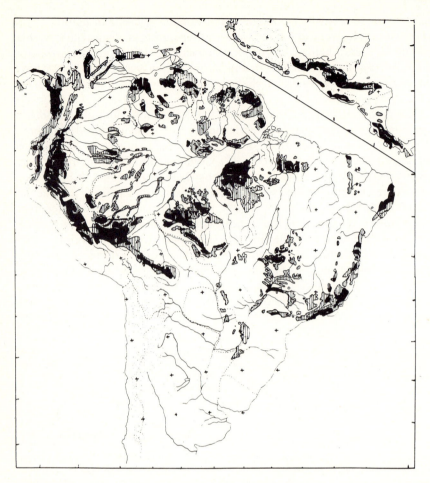

Fig. 14. Areas where tropical humid forest probably persisted in the last arid (Wisconsin-Würm) ice age deduced from all the different lines of geoscientific evidence. Sixty percent (*hatched*) and 80–100% (*black*) likelihood of persistence are shown based on overlap of positive evidence (After Brown 1987, Fig. 2.8). Blank areas may have been covered by relatively dry and more open forest types and/or by various kinds of nonforest vegetation. Gallery forests probably existed along many or most rivers in these latter regions.

alternating, prolonged periods of habitat continuity and habitat discontinuity during the Cenozoic. Rather, constantly changing climatic patterns may have caused continuous complex changes in the distribution of forest and nonforest elements during cold-arid, cold-humid, warm-arid, and warm-humid phases of the late Tertiary and Quaternary. Refuges may represent areas of relative habitat continuity and average survival of certain groups. During the height of at least some arid-glacial climatic phases, destruction of many fragmented but still identifiable community units (e.g., rain forest) may have continued in some regions of the world, e.g. in northern South America, to such an extent that very restricted populations of the endemic elements and many non-endemic plant and animal taxa persisted in dispersed and localized "mini-refugia" too small to register in the pollen record (Livingstone 1980).

   In contrast to lower Amazonia, no paleoecological information is yet available from the central and upper Amazonian forest region, as again pointed out by Connor (1986), Colinvaux (1987) and Salo (1987) in their critical discussions of the geoscientific basis of the refuge theory. Even if future surveys should suggest that the central and upper Amazon forest zone merely diminished in extent peripherally during arid climatic periods without fragmenting, the refuge theory remains a valid model for environmental forcing of evolution and the diversification of terrestrial faunas over vast regions of South America and other continents during the entire history of the earth. Cracraft and Prum (1988) emphasized the importance of paleogeographical changes for the differentiation of the neotropical biota and Räsänen et al. (1987) as well as Salo and Räsänen (1989) did so for the differentiation of extant species and subspecies in upper Amazonia. The latter authors assumed the isolation of animal populations on rather restricted uplifted arches separated by extensive flooded centers of sub-Andean sedimentary basins.

4 Discussion

The hierarchical disturbance processes affecting the neotropical forest region range from short-term gap-phase dynamics on a local scale to increasingly long-term regional and large-scale disturbance processes (paleoclimatic and tectonic disturbance). Each higher-level process in this hierarchy is superimposed upon the respective lower-level processes: Gap-phase dynamics occurs within a forested region affected by fluvial dynamics; the general region itself may be subject to climatic effects. Paleoclimatic and tectonic dynamics lead to large-scale changes in the distribution of forest and nonforest vegetation in regions where, on a smaller geographical scale, river dynamics and gap-phase dynamics regenerate the forest continuously. These disturbance processes are cyclic in the sense of a repetitive sequence of changes. All stages of the disturbance cycles are transitory, i.e. none represent a "beginning" or an "end phase". Individual cycles are not necessarily of the same duration. These processes are examples of "Time's cycle" (Gould 1987) without a component of direction. On the other hand, biotic differentiation (evolution) as a process of "Time's arrow" places these abiotic cyclic processes in a directional context.

   Evolutionary change at the level of subspecies and species in vertebrate and invertebrate groups may be affected differently by the various cyclic disturbance processes in the neotropical forest region. Whereas gap-phase dynamics may provide sufficient spatial separation in the case of small invertebrate populations for subspecies or species differentiation to occur, this is certainly not the case for vertebrates. Gap-phase dynamics provides habitat heterogeneity and thus helps to maintain the high tropical vertebrate species diversity but it is not a speciation factor, i.e., a causal factor for the origin of the high diversity of vertebrate species.

Similarly, river dynamics creates habitat heterogeneity and thus contributes to the maintenance of the high species diversity of invertebrate and vertebrate groups. Its role as a factor for the spatial separation of populations and ensuing speciation appears obvious in the case of invertebrate groups but is less certain in vertebrate animals. There is no question that the lower, broad portions of many Amazonian rivers, often in conjunction with their much wider floodplains, effectively separate populations of many forest birds, especially those species inhabiting the interior of the terra firme forest, thus inducing genetic differentiation of the separated populations. This has been pointed out by many authors over the last 100 years and has been demonstrated by Caparella (1988) on the basis of his study of allozymic differences among river-separated birds that do and do not show plumage differences. However, genetic differentiation is a necessary but not a sufficient stage in the speciation process. Differentiating populations must pass through a stage of complete geographic separation on all sides in order to possibly develop critical genetic isolation and thus attain species status.

The genetic differences between many populations separated by broad rivers disappear clinally toward the headwater region of the respective rivers where the latter are narrow and cease to be barriers. In the headwater regions, more or less uninhibited gene flow obliterates the differences observed between the populations occupying opposite banks of the lower river portions. Any discussion of the importance of rivers for the diversification of vertebrate faunas is incomplete without a consideration of the present and past ecological conditions in the headwater regions of rivers in which the wide, lower portions do indeed represent effective barriers to dispersal of many rain forest animals (River-refuge theory).

Paleoclimatic and tectonic disturbance processes probably caused massive regional separation of animal populations and were important factors in the differentiation of vertebrate faunas in the neotropics throughout the Cenozoic (Tertiary and Quaternary). During the Tertiary period, such disturbance processes occurred simultaneously with the tectonic processes, affecting large areas on a very long time scale (in the order of millions of years). Paleoclimatic dynamics on a scale of tens and hundreds of thousands of years increased worldwide in intensity during the late Miocene and Pliocene (toward the end of the Tertiary) and extended through the last 2 million years (Quaternary period, i.e., Pleistocene and Holocene) to the present time. It appears feasible, though by no means certain, that many extant species of forest birds and other animals originated within forest refuges during one of the dry climatic periods of the late Tertiary or early Pleistocene. Cracraft and Prum (1988) emphasized the effect of paleogeographic changes in explaining the origin of many presently existing neotropical bird species and their distribution patterns, although they conceded (1988, p. 617): "The refuge hypothesis remains a viable explanation for the patterns of vicariance documented" for many groups of forest birds. However, I am unable to follow their statement (p. 616) that a Pleistocene age of most species level taxa within the Amazonian biota ". . . is a logical consequence and critical component of the refuge hypothesis." Generally, this hypothesis postulates the isolation and differentiation of comparatively restricted animal populations in ecologically favorable refugia which formed during adverse climatic-vegetational phases of any historical period of the earth. In other words, refuge theory refers to the postulated origin of species in ecological refugia irrespective of the time period. As I stated previously (Haffer 1982, p. 6), "the Quaternary refuge theory does not propose that all extant species date from the Quaternary nor that all speciation has taken place in ecological refuges. Rather, it attempts to explain the latest and perhaps most effective of the series of differentiation events beginning during the Mesozoic [and continuing through the Tertiary] that contributed to the development of the modern biotas of the world" (see also Brown 1987).

5 Summary

The avifaunas of tree-fall gaps and the surrounding rain forest in the neotropical lowlands are quite distinct but their compositions overlap considerably because gaps and mature forest phases are end points on a local habitat gradient. Gap specialists among birds are predominantly frugivorous. Plant successional zones along meandering river courses in western Amazonia are sharply defined and lead to conspicuous habitat selection of birds. The most distinctive communities with the highest proportion of exclusive avian species inhabit the end points of this successional gradient (high-ground forest and *Tessaria* zone, respectively). The ecological processes maintaining the habitat heterogeneity and local mosaic distribution patterns of neotropical forest birds are cyclic regeneration of the forest mainly due to gap-phase dynamics and fluvial dynamics. These disturbance processes are cyclic in the sense of a repetitive sequence of changes. All stages of the disturbance cycles are transitory, i.e. none represent a "beginning" or an "end phase". Turnover rates are about 100 years for gap-phase dynamics to several 100 and 1000 to 2000 years in the case of fluvial dynamics. These rapid site turnover processes maintain the high tropical habitat diversity as a prerequisite for the maintenance of the high species diversity.

Numerous closely allied bird species replace each other geographically along sharply defined contact zones. The ranges of representative species form regional mosaic distribution patterns which often extend across large portions of the Neotropical Region. The origin of these mosaic patterns is not yet certain but they probably originated through historical disturbance processes such as paleoclimatic-vegetational cycles and/or paleogeographical changes in the distribution of land and sea areas including the flooding of wide river valleys in Amazonia. Individual cycles probably occurred at specifiable intervals or episodically at less regular intervals. These disturbance processes probably led to repeated fragmentations of the distributional areas of wide-ranging ancestral species, thus inducing various intensive differentiation and speciation of the separated representative populations.

The disturbance processes underlying the nested sets of mosaic distribution patterns of neotropical birds represent a hierarchy from local short term gap-phase dynamics to increasingly regional and long-term processes.

References

Amadon D (1966) The superspecies concept. Syst Zool 15:245–249

Bigarella JJ, Ferreira AMM (1985) Amazonian geology and the Pleistocene and the Cenozoic environments and paleoclimates. In: Prance GT, Lovejoy TE (eds) Amazonia. Key environments. Oxford, Pergamon Press, pp 49–71

Bock WJ, Farrand J Jr (1980) The number of species and genera of Recent birds: a contribution to comparative systematics. Am Mus Novitates 2703:1–29

Brokaw NVL (1985) Gap-phase regeneration in a tropical forest. Ecology 66:682–687

Brown KS Jr (1987) Areas where humid tropical forest probably persisted. Conclusions, synthesis, and alternative hypotheses. In: Whitmore TC, Prance GT (eds) Biogeography and Quaternary history in tropical America. Oxford, Clarendon Press, pp 44–45 and 175–196

Caparella AP (1988) Genetic variation in neotropical birds: implications for the speciation process. Acta XIXth Congr Int Ornith Ottawa 1986, 2:1658–1664

Colinvaux P (1987) Amazon diversity in light of the paleoecological record. Q Sci Rev 6:93–114

Connor EF (1986) The role of Pleistocene forest refugia in the evolution and biogeography of tropical biotas. TREE 1:165–168

Cracraft J, Prum RO (1988) Patterns and processes of diversification: speciation and historical congruence in some neotropical birds. Evolution 42:603–620

Delacour J, Amadon D (1973) Curassows and related birds. Am Mus Nat Hist, New York

Delcourt HR, Delcourt PA, Webb T III (1983) Dynamic plant ecology: the spectrum of vegetational change in space and time. Q Sci Rev 1:153–175

Di Castri F, Hadley M (1988) Enhancing the credibility of ecology: interacting along and across hierarchical scales. GeoJournal 17:5–35

Garwood NC, Janos DP, Brokaw N (1979) Earthquake-caused landslides: a major disturbance to tropical forests. Science 205:997–999

Goldammer JG, Seibert B (1989) Natural rain forest fires in eastern Borneo during the Pleistocene and Holocene. Naturwiss 76:518–520

Gould SJ (1987) Time's arrow and Time's cycle. Myth and metaphor in the discovery of geological time. Harvard Univ Press, Cambridge, Massachusetts

Haffer J (1969) Speciation in Amazonian forest birds. Science 165:131–137

Haffer J (1970) Art-Entstehung bei einigen Waldvögeln Amazoniens. J Ornith 111:285–331

Haffer J (1974) Avian speciation in tropical South America. Publ. Nuttall Ornith Club 14:390 pp

Haffer J (1977) A systematic review of the neotropical ground-cuckoos (Aves, *Neomorphus*). Bonner Zool Beitr 28:48–76

Haffer J (1982) General aspects of the refuge theory. In: Prance GT (ed) Biological diversification in the Tropics. Columbia Univ Press, New York, pp 6–24

Haffer J (1986) Superspecies and species limits in vertebrates Z Zool Syst Evol Forsch 24:169–190

Haffer J (1987a) Quaternary history of tropical America. In: Whitmore TC, Prance GT (eds) Biogeography and Quaternary history in tropical America. Clarendon Press, Oxford, pp 1–18

Haffer J (1987b) Biogeography of neotropical birds. In: Whitmore TC, Prance GT (eds) Biogeography and Quaternary history in tropical America. Clarendon Press, Oxford, pp 105–150

Hartshorne GS (1978) Treefalls and tropical forest dynamics. In: Tomlinson PB, Zimmermann MH (eds) Tropical trees as living systems. Cambridge Univ Press, Cambridge, pp 617–638

Hartshorne GS (1980) Neotropical forest dynamics. Biotropica 12 (Suppl):23–30

Levey DJ (1988a) Tropical wet forest treefall gaps and distributions of understory birds and plants. Ecology 69:1076–1089

Levey DJ (1988b) Spatial and temporal variation in Costa Rican fruit and fruit-eating bird abundance. Ecol Monogr 58:251–269

Livingstone DA (1980) History of the tropical rainforest. Paleobiology 6:243–244

Mayr E (1963) Animal species and evolution. Harvard Univ Press, Cambridge, Massachusetts

Mayr E, Short LL (1970) Species taxa of North American birds. Publ Nuttall Ornith Club 9, Cambridge, Mass, 127 pp

Oldeman RAA (1989) Dynamics in tropical rain forests. In: Holm-Nielsen LB, Nielsen IC, Balslev H (eds) Tropical forests. Botanical dynamics, speciation and diversity. Academic Press, London, pp 3–21

Paynter RA Jr (1972) Biology and evolution of the *Atlapetes schistaceus* species group (Aves: Emberizinae). Bull Mus Comp Zool 142:297-320

Paynter RA Jr (1978) Biology and evolution of the avian genus *Atlapetes* (Emberizinae). Bull Mus Comp Zool 148:323-369

Räsänen ME, Salo JS, Kalliola RJ (1987) Fluvial perturbance in the western Amazon basin: regulation by long-term sub-Andean tectonics. Science 238:1398-1401

Remsen JV, Parker TA III (1983) Contribution of river-created habitats to bird species richness in Amazonia. Biotropica 15:223-231

Richards PW (1952) The tropical rain forest. Cambridge Univ Press, Cambridge

Robinson SK, Terborgh J, Fitzpatrick JW (1988) Habitat selection and relative abundance of migrants in southeastern Peru. Acta XIXth Congr Int Ornith Ottawa 1986, 2:2298-2307

Rosenberg GH (1990) Habitat specialization and foraging behavior by birds of Amazonian river islands in northeastern Peru. Condor 92:427-443

Saldarriaga JG, West DC (1986) Holocene fires in the northern Amazon basin. Q Res 26:358-366

Salo J (1987) Pleistocene refuges in the Amazon: evaluation of the biostratigraphical, lithostrati- graphical and geomorphological data. Ann Zool Fennici 24:203-211

Salo J, Kalliola RJ (1989) River dynamics and natural forest regeneration in Peruvian Amazonia. In: Jeffers J (ed) Rainforest regeneration and management. MAB(UNESCO), Book Ser, UNESCO and Cambridge Univ Press, Paris Cambridge (in press)

Salo J, Räsänen M (1989) Hierarchy of landscape patterns in western Amazon. In: Holm-Nielsen LB, Nielsen IC, Balslev H (eds) Tropical forests. Botanical dynamics, speciation and diversity. Academic Press, Lond New York, pp 35-45

Salo J, Kalliola R, Häkkinen I, Mäkinen Y, Niemelä P, Puhakka M, Coley PD (1986) River dynamics and the diversity of Amazon lowland forest. Nature (Lond) 322:254-358

Sanford RL Jr, Saldarriaga J, Clark KE, Uhl C, Herrera R (1985) Amazon rain-forest fires. Science 227:53-55

Schemske DW, Brokaw N (1981) Treefalls and the distribution of understory birds in a tropical forest. Ecology 62:938-945

Schubert C (1988) Climatic change during the last glacial maximum in northern South America and the Caribbean: a review. Interciencia 13:128-137

Snethlage E (1913) Über die Verbreitung der Vogelarten in Unteramazonien. J Ornith 61:469-539

Snow DW (1982) The cotingas: Bellbirds, umbrellabirds and their allies. Brit Mus Nat Hist, Oxford Univ Press, Lond

Stiles FG (1983) Birds. In: Janzen DH (ed) Costa Rican Natural History. Chicago Univ Press, Chicago, pp 502-544

Terborgh J (1985) Habitat selection in Amazonian birds. In: Cody M (ed) Habitat selection in birds. Academic Press, Lond New York, pp 311-338

Terborgh J, Robinson SK, Parker TA III, Munn CA, Pierpont N (1990) Structure and organization of an Amazonian forest bird community. (in press)

Tricart J (1985) Evidence of Upper Pleistocene dry climates in northern South America. In: Douglas I, Spencer T (eds) Environmental change and tropical geomorphology. Allen & Unwin, Lond, pp 197-217

Uhl C, Clark K, Dezzeo N, Maquirino P (1988) Vegetation dynamics in Amazonian treefall gaps. Ecology 69:751-763

Whitmore TC (1989) Canopy gaps and the two major groups of forest trees. Ecology 70:536-538

Willis EO (1969) On the behavior of five species of *Rhegmatorhina*, ant-following antbirds of the Amazon basin. Wilson Bull 81:363-395

Willis EO (1983) Toucans (Ramphastidae) and hornbills (Bucerotidae) as ant followers. Le Gerfaut 73:239-242

Small Mammals and the Mosaic-Cycle Concept of Ecosystems

H. Korn[1]

1 Introduction

For many decades there have been reports that small mammals in grassland ecosystems destroy substantial parts of their habitat (e.g., Merriam 1902; Dusen 1903; Przewalski in Formozov 1928, 1978). Likewise foresters know that small mammals have a great potential to destroy seed and that they are able to destroy entire seed crops (Sullivan 1979). They can also kill extensive amounts of seedlings and due to both processes many reforestation efforts are unsuccessful (e.g., Smith and Aldous 1947; Hooven 1971; Crouch 1986). From man's point of view, small mammals, especially rodents, were therefore thought to be evil and were destroyed whenever possible. Only recently there is growing appreciation that small mammals can play a key role in the structure and function of natural ecosystems (e.g., Huntly and Inouye 1988; Korn and Korn 1989).

Current scientific literature was surveyed in regard to the effects of small mammals on soils, vegetation, and other animals in grasslands, woodlands, arid zones, and Arctic tundra. Areas with high levels of agricultural activity were generally excluded from the analysis since plant succession lasts only a short time. Rangelands are included since natural grasslands are nowadays rare and virtually unstudied with respect to rodent disturbance. Since the soil structure is unaltered, the difference between grazing of livestock and naturally occurring grazers (bison, antelope, etc.) is not believed to be crucial for the functioning of the ecosystem. In contrast, intensively used meadows, mainly in Europe, were not considered since it is well known that high densities of large domestic mammals can severely change plant communities by selective feeding and trampling. These systems are also heavily fertilized and natural cycles are interrupted. Data on managed or secondary forests are included since information about the ecology of small mammals in virgin forests is very rare.

For convenience the species up to the size of marmots and badgers are considered in the analysis, since their effect on the ecosystem differs only in quantity but not in quality from that of smaller, burrowing mammals.

The aim of this investigation is to link the effects of small mammals on soils, plants, and other animals to the mosaic-cycle concept of ecosystems. For a review of the concept, see Remmert (1985, 1988).

[1] Programa Regional de Vida Silvestre para Mesomerica y el Caribe, Escuela de Ciencia Ambinentales, Universidad Nacional, Apartado 1350, Heredia, Costa Rica

H. Remmert (Ed.)
Ecological Studies Vol. 85
© Springer-Verlag Berlin Heidelberg 1991

2 Effects of Small Mammals on Low Vegetation Ecosystems: Grasslands, Tundra, Subtropical and Tropical Arid Zones

These ecosystems are characterized by low vegetation. They will be treated together because often the boundaries between them are difficult to define. Grasslands are comparatively simple in their vegetation structure and easily accessible. Therefore, they are the best-studied ecosystems in terms of small mammal disturbances. Additionally, many small mammal species live in colonies or high density patches, e.g. prairie dogs. Their individual influences on soil and vegetation are therefore multiplied and especially obvious Fig. 1. One reason for their coloniality may even be that they can change their habitat as they like. This normally means that the vegetation in active small mammal colonies is lower than in surrounding areas (Dusen 1903; Löffler and Margules 1980; Agnev et al. 1986; Whicker and Detling 1988; Korn and Korn 1989), presumably to facilitate predator detection (King 1955). Because grasslands and other low vegetation ecosystems are comparatively poor in cover, most mammal species smaller than 10 kg live in burrows at least part of the time for protection from predators and adverse climatic conditions. Some rodent families (e.g., Geomyidae in North and Central America, Ctenomyidae in South America, Bathyergidae in Africa south of the Sahara, and Spalacidae in southeast Europe, west Asia and north Africa) even live almost completely under ground where they also have their feeding grounds. The same can be said for many insectivorous and marsupial moles (Chrysochloridae, Talpidae, Noctoryidae). When they excavate their burrows a large amount of surplus soil has to be brought up to the surface, thus causing disturbance of the vegetation pattern (Figs. 2–5). The size of such a disturbance, which is additionally enhanced when the animal selectively

Fig. 1. Prairie dog (*Cynomys ludovicianus*) in front of its burrow. Note the differences in vegetation caused by the activity of the rodent (Photo H. Remmert)

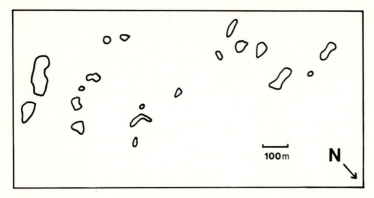

Fig. 2. Distribution and extent of gerbil (*Tatera* sp.) colonies in an area of the Nylsvley savanna. Map drawn after an aerial photo taken in April 1974

Fig. 3. *Tatera brantsii*, a colonial rodent that vigorously digs in areas of sandy soil in the Nylsvley Nature Reserve, South Africa

Fig. 4. Soil heaps in front of gerbil (*Tatera brantsii*) burrows, covering vegetation in the Nylsvley Nature Reserve. Dry season aspect. August 1986

Fig. 5. View over the Nylsvley savanna with an active gerbil colony (*Tatera brantsii*) in the foreground, August 1986

feeds from the vegetation or collects nesting material, may range from about 100 cm^2 on the surface made by porcupines (Gutterman and Herr 1981) or may extend to more than 100 ha, e.g., large prairie dog towns (Osborn and Allan 1949; Campbell and Clark 1981; Dallsted et al. 1981), wombat colonies (Löffler and Margules 1980) or *Ctenomys* sites (Dusen 1903). The reported maximum was a single prairie dog colony that occupied about 65 km^2 in Texas (Merriam 1902) (Table 1). In extreme cases almost 40% of the area can be affected by rodent burrows, but generally the area of disturbance is less (Table 2). The amount of soil brought up to the surface in such areas can be far greater than 100 t ha^{-1} a^{-1} but is normally lower (Table 3). The effects on the soil and vegetation that can be observed depend on the size of the pile or heap, the quality of the soil excavated, and the thickness of the soil covering the plants. Another influence of small mammals on patch dynamics may also be due to selective predation, hoarding, and movement of seeds. This topic was recently reviewed by Wiens (1985) for desert environments but the focus was almost exclusively on seed preferences and microhabitat selection of small rodents. Three groups of plants are generally well adapted to disturbances caused by burrowing mammals: (1) the hemicryptophytes, which can easily grow through thicker layers of soil (Laycock 1958; Zhuchkova and Utekhin 1975; Skoczen et al. 1983) as long as the nutritious roots or rhizomes are not consumed by rodents; (2) plants that can rapidly invade open space with rhizomes (Ellison and Aldous 1952; Skoczen et al. 1983); and (3) therophytes the seeds of which are already present or can rapidly invade open sites. They can make rapid use of this competition-free zone (Laycock 1958; Skoczen et al. 1983). Positive effects of the small mammals on the system are that the soil is loosened and the ventilation of air is increased (Ellison and Aldous 1952; Berman et al. 1966; Hansen and Morris 1968). The penetration of rainwater is facilitated (Greene and Reynard 1932; Ingles 1952; Berman et al. 1966; Weiner et al. 1982; Cox 1987) but, on the other hand, in some cases erosion can also occur due to the loose soil on the surface (Formozov 1928; Ellison 1946). In contrast, Weiner et al. (1982) report that rodent activity in central Mongolia restricts the erosion process due to stimulation of primary production, acceleration of element cycling, and retardation of water evaporation. Moreover, the microflora and the topsoil become mixed and the mineral composition of the soil is changed (Formozov 1928; Greene and Murphy 1932; Greene and Reynard 1932; Hansen and Morris 1968; Chesemore 1969; Abaturov 1972; Thompson 1972; McDonough 1974; Laycock and Richardson 1975; Mielke 1977; Weiner et al. 1982; Spencer et al. 1985; Huntly and Inouye 1988). Small mammals also increase the organic compounds in the soil due to food remains, nesting material, buried vegetation, feces, and carcasses (Greene and Reynard 1932; Taylor 1935; Ellison and Aldous 1952; Ingles 1952; Berman et al. 1966; Zhuchkova and Utekhin 1975; McKendric et al. 1980; Grant and McBrayer 1981; Weiner et al. 1982; Zielinski 1982; Schauer 1987a). All these mentioned effects bring about quantitative and qualitative alterations of the plant cover. The changes of the vegetation which can actually be observed depend on the age of the disturbance, its intensity, its size, the prevailing vegetation, the season, and the type of soil brought to the surface (Osborn and Allan 1949; Bonham and Lerwick 1976; Hobbs and Mooney 1985;

Table 1. Surface area of individual disturbances caused by burrowing mammals

Habitat and location	Species	Size of disturbance	Type of disturbance	Authority
Pastures in Poland	Talpa europaea	0.14 m²	Mole hill	Skoczen et al. (1983)
Steppe in Central Asia	Marmota bobak	221 m²	Mound at burrow entrance	Jettmar (1926)
Semidesert, USSR	Citellus pygmaeus	8.3 m²	Mound at burrow entrance	Mushketov in Formozov (1928)
Desert steppe in China	Citellus dauricus	0.04 m²	Blank area at burrow entrance	Xia and Zhong (1978)
Desert mountains in Israel	Hystrix indica	0.01 m²	Diggings for bulbs	Gutterman and Herr (1981)
Steppe in northern Mongolia	Myospalax aspalax	0.125 m²	Soil heap	Schauer (1987b)
Arid steppe in Mongolia	Microtus brandti	174.3 m²	Area of colony	Zielinski (1982)
High arctic tundra, Canada	Dicrostonyx groenlandicus	11.8 m²	Lemming mats	Mallory and Boots (1983)
Short grasslands, Tanzania	Tatera robusta	0.2 m²	Soil heap	Senzota (1984)
Savanna, South Africa	Tatera brantsii	0.5–0.9 ha	Area of colony	Korn (unpubl.)
Namib desert, Namibia	Gerbillus sp.	27.3 m²	Gerbil colonies	Cox (1987)
Bunchgrass, southern Texas	Geomys attwateri	0.15–0.18 m²	Gopher mound	Spencer et al. (1985)
Old field, Minnesota	Geomys bursarius	0.18 m²	Gopher mound	Tilman (1983)
Shortgrass prairie, Colorado	Thomomys talpoides	0.12 m²	Gopher mound	Grant et al. (1980)
Aspen range, Utah	Thomomys talpoides	0.14 m²	Gopher mound	McDonough (1974)
Grassland, California	Thomomys bottae	0.13 m²	Gopher mound	Koide et al. (1987)
Grassland, Texas	Cynomys sp.	65 km²	Colony	Merriam (1902)
Prairie in Wyoming	Cynomys leucurus	43 ha (2–184 ha)	Colony	Campbell and Clark (1981)
Rolling plains, Wyoming	Cynomys ludovicianus	35 ha (1–189 ha)	Colony	Campbell and Clark (1981)
Prairie, South Dakota	Cynomys ludovicianus	5 –250 ha	Colony	Dahlsted et al. (1981)
Prairie, South Dakota	Cynomys ludovicianus	0.8–3.1 m²	Mound at burrow entrance	Whicker and Detling (1988)
Tallgrass prairie, Iowa	Taxidea taxus	0.2–0.3 m²	Diggings of ground squirrels	Aikman and Thorne (1956)
Arctic tundra, Alaska	Alopexlagopus innuitus	about 30 m²	Den site	Chesemore (1969)

Table 2. Area of disturbance due to burrowing activity of small mammals

Habitat and location	Species	Disturbance area	Authority
Grassland in Germany	*Talpa europaea*	0 – 5.4%	Knapp (1959)
Grassland in Poland	*Talpa europaea*	0 –22.4%	Skoczen et al. (1983)
Coastal dunes, Europe	*Oryctolagus cuniculus*	0.6– 1.3%	Burggraf and Meijden (1984)
Puna in Andes of Argentina	*Ctenomys* (frater)	Small	Werner (1977)
Grasslands in Tanzania	*Tatera robusta*	2%	Senzota (1984)
Steppe, northern Mongolia	*Myospalax aspalax*	12.8%	Schauer (1987b)
Steppe near Voronej, USSR	*Spalax microphthalmus*	0.2 and 0.5%	Obolensky in Formozov (1928)
Desert steppe, China	*Citellus dauricus*	0.04%	Xia and Zhong (1978)
Steppe near Saratov, USSR	*Citellus pygmaeus*	2%	Dimo and Keller in Formozov (1928)
Arid steppe in Mongolia	*Microtus brandti*	12%	Weiner and Gorecki (1982)
Arid steppe in Mongolia	*Microtus brandti*	37.6%	Zielinski (1982)
Steppe in USSR	*Marmota* sp.	10%	Glinka in Formozov (1928)
Saline steppe in USSR	*Lagurus lagurus*	6 –10%	Abaturov (1964)
Forest plantations in Oregon	*Thomomys monticola*	>25%	Hooven (1971)
Aspen woodland in Utah	*Thomomys talpoides*	6 –22%	McDonough (1974)
Shortgrass prairie, Colorado	*Thomomys talpoides*	2.5–8%	Grant et al. (1980)
Subalpine zone in Utah	*Thomomys talpoides*	3.5%	Ellison (1946)
Subalpine grassland in Utah	"Pocket gopher"	11%	Laycock and Richardson (1975)
Mountain rangeland, Colorado	"Pocket gopher"	18%	Turner et al. (1973)
Old field, Minnesota	*Geomys bursarius*	2.2%	Tilman (1983)
Oak and pine forest, Florida	*Geomys pinetus*	4%	Kalisz and Stone (1984)
Bunchgrass, south Texas	*Geomys attwateri*	9.4 and 9.9%	Spencer et al. (1985)
Rangeland, Texas	*Geomys breviceps*	8%	Buechner (1942)
Prairie in South Dakota	*Cynomys ludovicianus*	12%	Whicker and Detling (1988)
Rolling plains, Wyoming	*Cynomys ludovicianus*	0.7%	Campbell and Clark (1981)
Prairie in South Dakota	*Cynomys leucurus*	3.2%	Campbell and Clark (1981)
Desert Rangeland, New Mexico	*Dipodomys spectabilis*	2%	Moroka et al. (1982)
"Fynbos", South Africa	*Bathyergidae*	28.2%	Reichman and Jervis (1989)

Inouye et al. 1987; Koide et al. 1987). In a fresh rodent, lagomorph, or marsupial colony vegetative cover and plant biomass are generally decreased (e.g., Przewalsky in Formozov 1928; Löffler and Margules 1980; Agnev et al. 1986; Korn and Korn 1989), but when the colony ages or deteriorates a succession starts (Osborn and Allan 1949; Foster and Stubbendieck 1980) which may lead to increased biomass and productivity of the site (Ellison and Aldous 1952; Ingles 1952; Thompson 1972; Zielinski 1982; Fig. 6). This eventually leads to an increased productivity of the ecosystem as a whole (Laycock and Richardson

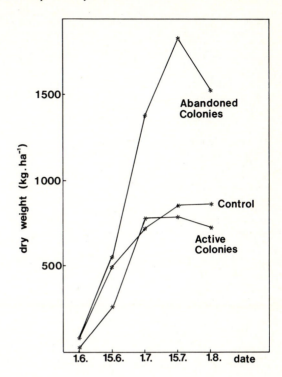

Fig. 6. The growth rate of organic matter in plants covering the arid steppe of central eastern Mongolia (After Zielinski 1982)

1975; Kiryushchenko 1978; Grant et al. 1980; Weiner et al. 1982). Similar successions like in small mammal colonies, only on a much smaller scale, can be observed in single mole hills or gopher mounds. The vegetation in the early state usually differs to the original state, except when the disturbance is very small and neighboring plants grow in the open space (Burggraaf and van der Meijden 1984). Open sites are often excellent germination sites for certain plants that are present in lower densities or do not exist in the "background" vegetation (Zonov et al. 1983). Some plant species may even depend on such disturbances (McDonough 1974; Platt 1975; Werner 1977; Schaal and Leverich 1982; Leverich 1983; Zonov et al. 1983; Korn and Korn 1989) and may have coevolved with the small mammals. Generally, annual plants are favored by burrowing and other small mammal activities (Osborn and Allan 1949; McDonough 1974; Laycock and Richardson 1975; Foster and Stubbendieck 1980; Moroka et al. 1982; Schaal and Leverich 1982; Burggraaf and van der Meijden 1984; Cox 1987), but occasionally good germination sites for shrubs and trees are provided (Tevis 1956a; Rodin 1961; Zaletayev 1976; Walter and Breckle 1986). In many ecosystems the floristic composition is enriched due to small mammal activity and the species diversity of plants is increased (Kiryushchenko 1978; Williams and Cameron 1986; Inouye et al. 1987; Korn and Korn 1989).

All these effects, though unequally documented, seem to be general phenomena that can be observed worldwide. To link these "disturbances" to the mosaic-cycle concept of ecosystems (Remmert 1985, 1988) they need to be patchy and appear over time at different sites.

Table 3. A mound of soil worked up to the surface by small mammals

Habitat and location	Species	Amount of soil brought up to surface (in t/ha/a)[a]	Authority
Meadow in Czechoslovakia	*Talpa europaea*	55	Grulich (1959)
Spruce forest in USSR	*Talpa europaea*	3.9–18.6	Abaturov (1972)
Sand dune in desert, India	*Meriones hurrianae*	381*	Sharma and Joshi (1975)
Semidesert in USSR	*Citellus pygmaeus*	1.5	Abaturov (1972)
Steppe in Mongolia	*Myospalax aspalax*	10	Schauer (1987b)
Steppe in Mongolia	*Ellobius talpinus*	1.2	Schauer (1987b)
Semiarid areas, Argentina	*Ctenomys* sp.	30	Contreras and Maceiras (1970)
Yosemity National Park, California	"Pocket gophers"	20	Grinnell (1923)
Subalpine rangeland, Utah	*Thomomys talpoides*	12.5	Ellison (1946)
Shortgrass prairie, Colorado	*Thomomys talpoides*	11	Grant et al. (1980)
Vegetable garden, Utah	*Thomomys talpoides*	95**	Richens (1966)
Mountain meadow, California	*Thomomys monticola*	18.5	Ingles (1952)
Alfalfa fields, California	*Thomomys bottae*	max. 117	Miller (1957)
Pine forest, Florida	*Geomys pinetus*	0.3	Kalisz and Stone (1984)
Pine forest, Florida	*Geomys pinetus*	8.2	Kalisz and Stone (1984)
Overgrazed range, Texas	*Geomys breviceps*	17.5	Buechner (1942)
Ungrazed range, Texas	*Geomys breviceps*	0.9	Buechner (1942)
Savanna in South Africa	*Tatera brantsii*	16.9***	Korn (unpubl.)

[a] *Calculated from daily rates; ** calculated from monthly rates per animal with a density of 75/ha; *** calculated from data taken in dry season only; inside colony.

The oldest examples that support the validity of the concept come from Russian literature, starting in the last century with the work of Przewalsky. Extensive work has been done since in Eurasia.

Formozov (1928) summarized the effects of rodents on Asian steppe. The vegetation on the mounds produced by the numerous small mammals in the midst of the steppe usually has a more desert, xerophytic character. At first it contains many "weeds". The earth deposited on the surface undergoes changes analogous to those which the soil of the surrounding steppe has passed through earlier.

Simultaneously with the process of soil formation a change of vegetation takes place on the hillocks. Finally, the vegetation becomes similar to the areas that have not been unearthed. In the steppe occupied by rodents, neither the superficial layers nor the vegetation are homogeneous, but varied. Through the digging activity of the rodents, portions of the steppe continually revert to their original stage of development. The rodents postpone the natural succession of plant associations, and thereby assist in maintaining for long periods the kind of vegetation to which they have become adapted.

Kozhemyakin (1978) writes that in the desert steppes of Mongolia the activity of *Citellus erythrogenys* makes a marked impact upon vegetation cover. With diverse age of burrows specific changes in the vegetation take place. The rodent activity leads to the intensification of the vegetation pattern. Another example is given by Stogova (1986). In mountain pasture in the Kazakh SSR, USSR, mole voles transformed the soil, thus contributing to microrelief and microphytocenosis formation. The soil structure changed. Soil and vegetation became mosaic, which manifested itself in the diversity of pastures. Hsia and Chong

(1966) observed that in abandoned fields in the desert steppe of Inner Mongolia, due to the activity of gerbils (*Meriones unquiculatus*), a mosaic pattern of the vegetation had formed.

Przewalsky described the interaction between a small lagomorph and its environment. Where the pika (*Ochotona melanostoma*) settle, they devour the grass and its roots completely, digging them out of the earth, so that extensive surfaces of meadow in the Koukou-Nor, as well as in northern Tibet, become entirely waste. Then the small animals migrate to areas in the vicinity, while the surface from which they have devoured everything, regenerates again little by little and, probably, in time will be again occupied by pikas (Przewalsky in Formozov 1928).

Walter and Breckle (1986) report in a review that in the same ecosystem small rodents (*Microtus* sp., *Lagurus lagurus*) dig only through the superficial layers of soil but do not effect it to a great extent. Instead they graze intensively around their burrow until almost no plant matter remains, not even the underground parts of tuft grasses. When the soil is completely denuded by grazing, the colony moves to another area. At the site which is fertilized with urine and feces, a repopulation starts with nitrophile plant species followed by a succession of perennial species until the original vegetation is restored. At the same site Lavrenko (1952) observed that the burrows of rodents (*Citellus* sp., *Marmota* sp.) are very abundant and regularly moved, resulting in a micromosaic of the vegetation in semiarid steppes of Eurasia. Moreover, the plant cover of the steppe is not as homogeneous as it appears to the superficial observer. Due to the activity of the rodents a rotation of the microsites originates which means that the plant species alternate on single sites. This is probably the prerequisite for a continuous conservation of zonal vegetation with an unchanged composition on larger areas (Walter and Breckle 1986).

Another continent with many examples that support the mosaic-cycle concept is North America. The most important animals in this respect are gophers and prairie dogs. Gopher mounds represent a mosaic of sites, constantly shifting in time and space, which exist individually in one of several soil surface conditions: bare, invaded and dominated by annuals with a few seedlings of perennials, and dominated by perennials with a few persistent annuals in peripheral areas (McDonough 1974). The cycle is estimated to be completed within a 4-year period (Fig. 7). Laycock (1958) observed that where the pocket gopher is part of the biotic community the gopher mounds are microsites where pioneer species (therophytes) are continually perpetuated. These stable or "climax" communities will always include the pioneer species. Williams et al. (1986) report that in Texas coastal prairie species diversity and biomass of dicots may be maintained in a mixed forb-grassland community by the continuous, but spatially heterogeneous, disturbance by pocket gopher herbivory and deposition of mounds. Hobbs and Mooney (1985) suggest that continued gopher disturbance is a factor allowing several species, including perennial grasses, to persist in the community. It was estimated that gophers can turn over as much as 30% of the total area per annum.

According to Grant et al. (1980), in short grass prairie ecosystems in Colorado the plant community might be viewed as a mosaic of gopher mounds in various

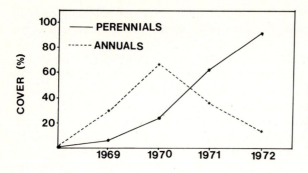

Fig. 7. Percentage of annual and perennial plant species cover over 4 years of colonization after disturbance by pocket gophers (After McDonough 1974)

stages of secondary succession. Reichman and Smith (1985) also mention that specific impacts of gophers (*Geomys bursarius*) at any one time could become controlling factors, as, over many years, pocket gophers burrow through an area, each time exacting their toll and resetting the successional schedule or nudging the community in another direction. The array of impacts, both detrimental and benificial, should create a mosaic of successional microsites.

Mohr and Mohr (1936) stated that inhabitable but unoccupied ground between groups is more obvious in the case of pocket gophers, prairie dogs, and similar colonies than among wider ranging animals. Such ground, which is called resting ground, is not to be confused with blanks or relatively uninhabitable areas. It is, on the other hand, ground which is vacated for a short time but likely to be used again by the same individual which vacated it, by neighboring individuals, or by their immediate offspring. In the case of pocket gophers, such areas begin as thoroughly burrowed and relatively denuded ground marked by weathered mounds. In the case of prairie dogs, *Cynomys gunnisoni*, resting areas begin as deserted "towns". Taylor (1924) reported that towns are occupied continuously until the vegetation is used up. The margins of these areas usually show a series of prairie dog towns gradually encroaching upon the untouched grassland.

Smith (1967) revisited a prairie dog town after 10 years. He wrote that the town had expanded eastward 150 ft. in a strip about 300 ft. long from north to south and the prairie dogs had withdrawn from an area of corresponding size on the northwestern part of the dog town. The total area of the town was about the same size as it was 10 years before.

For the southern hemisphere with the continents of South America, Africa, and Australia comparatively little published information supporting the mosaic-cycle concept is available. Nevertheless, there are examples for each continent. Reig (1970) reports that the fossorial rodent *Spalacopus cyanus* from Chile lives in nomadic colonies. The rodent species, in contrast to *Ctenomys*, seems to feed entirely under ground on succulent subterranean stems and tubers of *Leucocoryne ixioides* (Liliaceae). When the food supply of an area is depleted, the colony moves on into new areas. Within an active colony there is usually one burrow opening with a soil mound per square meter. The effect on the vegetation was obvious but not quantified.

Werner (1977) describes that tuco-tucos (*Ctenomys* sp.) in the Puna of Argentina move on after one area is bare of vegetation. The flora can then slowly regenerate on the disturbed sites. Effects of the gerbil *Tatera brantsii* on savanna vegetation in South Africa were intensively studied by Korn and Korn (1989). In different years the gerbil colony had its location at different sites (Fig. 8). They concluded that the gerbils are considered as an integral part of the system and necessary for the maintenance of a high species diversity of plants and animals. The regular "disturbance" of savanna vegetation followed by the "recovery" of the system is driven by the movement of gerbil colonies across the area.

In Australia a large marsupial, the hairy nosed wombat, *Lasiorhinus latifrons*, occurs in great numbers on the Nullabor Plain. Wombats tend to live in colonies. The burrowing and mound building of the animals create areas of bare ground and freshly dug soil. The wombats appear to graze near their warrens in expanding circles (Loeffler and Margules 1980). The animal will certainly move their warrens when distances between the den and the feeding sites become to large, allowing the vegetation to recover again.

The changes which can actually be observed due to small mammal disturbances, i.e., whether the soil brought up to the surface is good or bad, productivity is increased or decreased, etc., is not important for the general applicability of the mosaic-cycle concept in low vegetation ecosystems. It is only important that the disturbance differs from the surrounding area and undergoes a succession that eventually leads to the same status as before. The result is a mosaic pattern of vegetation patches with a different disturbance history that rotates in space over time.

Regarding the size of small mammal disturbances, we have quite good knowledge for many species (Table 1). The only problem is that different researchers have focused on different levels of disturbance. It is quite different to observe changes in individual piles, mole hills, and gopher mounds, or to study an entire gerbil or prairie dog colony. The latter can be considered as assemblages

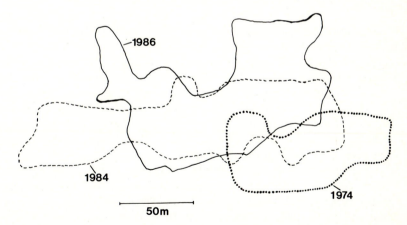

Fig. 8. Location of a large gerbil colony (*Tatera brantsii*) in three different years (Korn and Korn 1989)

of many smaller disturbances. In colonial rodents it is therefore possible to see the
cycles proceeding on two different levels. The smaller level contains individual
burrows which form, age, and deteriorate. New ones are constructed at a different
site within the colony. The second level is the entire colony, or burrow system,
which undergoes a successive cycle or even proceeds into virgin areas after the
original site has been depleted (Smith 1967; Reig 1970; Werner 1977; Korn and
Korn 1989).

The time span for the completion of a cycle is the least known factor in the
concept. The only available information exists from McDonough (1974), who
estimates the cycle of gopher mounds on aspen rangeland to be completed within
a 4-year period (Fig. 7). Under moist and warm conditions the process may be
quicker and under dry and cold conditions slower. Other disturbances, especially
large colonies of prairie dogs (*Cynomys* sp.) or great gerbils (*Rhombomys opimus*)
may persist for several centuries at the same site (Naumov and Lobachev 1975)
but will also eventually become extinct. The recovery of such a site may take
several decades or even centuries, e.g. a prairie dog town abandoned for 5 years
did not show major signs of recovery (Klatt and Hein 1978). Soil mounds at
burrow entrances represent small patches within a colony that have physiological
and chemical properties that may remain altered for hundreds or thousands of
years (Carlson and White 1987).

Another important question is, how large must an area be in order to secure
the persistence of these natural cycles for the future? So far, we do not have an
answer, not even a good guess to this problem, which is made even more
complicated by the severe population changes that many small mammal species
exhibit. Nevertheless, the low vegetation ecosystems will not be the same without
these small to medium disturbances caused by burrowing mammals. In the long
run they will certainly change their plant species composition and general
appearance.

3 Effects of Small Mammals on Forest Ecosystems

In contrast to changes induced by the activity of small to medium-sized mammals
in low vegetation ecosystems, their effects on mature forest ecosystems are
negligible. The densities of small mammals are generally less in climax forest than
earlier successional stages (e.g., Martell 1983; Teipner et al. 1983; Herrchen
1989). A possible impact of small mammals therefore concentrates on the
regeneration phase of the forest. Small mammals, especially rodents and
lagomorphs, eat seeds and seedlings and destroy small trees due to bark stripping
and root cutting (Capp 1976). On the other hand, they can also be very important
and necessary vectors for seed dispersal.

Since small mammal communities and their successional changes in natural
forest ecosystems are virtually unstudied, we have to look at managed forests to
find possible parallels from which we can extrapolate. After clear-cutting of an
area, changes in species composition and density of small mammals are observed
(Tevis 1956b; Ahlgren 1966; Hooven 1971; Gashwiler 1970; Capp 1976; Bor-

recco and Anderson 1980; Teipner et al. 1983). Gophers, for example, have the lowest densities in a climax forest. Their densities increase on lands disturbed by fire, logging, site preparation, or other events that open the canopy because their preferred food, early successional forbs and grasses, often thrive in these areas (for review, see Teipner et al. 1983). This often causes severe problems with reforestations (Smith and Aldous 1947; Hooven 1971; Borrecco 1976; Crouch 1982).

If there is a similar rise in density after natural disturbances, like storm and fire, which has been observed in gophers (Teipner et al. 1983) and European woodland rodents (Herrchen 1989), plant succession and plant species composition may be heavily influenced by small rodents. Many small rodents preferably eat seeds and seedlings of certain trees (e.g., Baeumler and Hohenadl 1980). Therefore, these species may be excluded from early successional stages of forest ecosystems due to seed predation. Only years later, when the canopy has closed and densities of certain rodent species are lower again, some tree species may have a chance to escape seed predation by small mammals. It is probably not a coincidence that in a mature Ohio forest where less than 15% of the reproductive trees were hickory and beech, with oak accounting for 67%, more than 50% of the seed diet of fox and gray sqirrels came from the first two species (Nixon et al. 1968). Baeumler and Hohenadl (1980) also reported a strong influence of rodents on the regeneration of trees, especially beech (*Fagus sylvatica*) in a mountain forest in southern Germany. Sullivan (1979) found that only a few deer mice (*Peromyscus maniculatus*) were required to destroy most of the conifer seeds dispersed over a clear-cut area in a few days. He stated that it is impossible to remove all animals, yet for the past 50 years, baiting with poison has been continually attempted, with very little success. Hooven (1971) came to the same conclusion for pocket gophers. Control of them, where it is related to reforestation, has generally been unsuccessful. Trapping or baiting is expensive and ineffective. The presence of a few pocket gophers per acre can drastically reduce the success of a *ponderosa* pine plantation.

Future research should therefore focus on the various survival chances of different seed types in relation to different population densities of small mammals. There should also be investigations on how different forest management practices influence population densities of small mammals and how different habitat alterations relate to habitat requirements of small mammals. The results should lead to a management form in which rodents do not play such an important role in reforestations.

Let us now look at the other phases of the forest cycle. The forest undergoes different stages with different species compositions in different parts of the world (Remmert 1985, 1988 and literature cited therein). Small mammal communities and population densities are expected to change likewise with the different plant communities in the cycle. Data for natural systems are rare and for no system is there a documentation of the entire cycle. From managed forest we know that small mammal communities change with the age of the stand (e.g., Atkeson and Johnson 1979; Martell 1983; Ramirez and Hornocker 1981; Jensen 1984) and that there are also differences between forest types (McKeever 1961; Ramirez and

Hornocker 1981). Nevertheless, it is beyond the scope of this review to present an extensive list of examples since the species combinations change from place to place, depending on the geographic ranges and local distributions of the small mammal species.

In a natural system all stages of the forest cycle are represented beside each other (Remmert 1988). Therefore, a parallel study of the different vegetation types should give a good indication of the changes in small mammal communities within the succession. To complete the cycle in a forest ecosystem several hundred to a few thousand years are needed (Remmert 1985, 1988) but within that time period small mammals seem to influence the system actively only within the very short time period of regeneration. The research of foresters and other investigators has therefore focused extensively on that topic. Within the other phases small mammal populations and communities seem to passively follow the changes in the environment. Nevertheless, there are indications that small rodents in woodland ecosystems are important agents of soil formation like in low vegetation ecosystems. Their holes favor the ventilation of root systems and facilitate the penetration of rainwater. Some rodents also live in holes in the base of old trees or use hollow trunks and branches as nesting sites. Their excrements, food, and litter remains favor the decomposition of the wood, finally leading to the death of branches or entire trees (for review of these topics, see Walter and Breckle 1983). Rodents also seem to be important agents in the dispersal of subterranean fungi and favor germination of spores (Fogel and Trappe 1978; Maser et al. 1978a,b; Kotter and Farentinos 1984; Baeumler 1986; Blaschke and Baeumler 1986, Ovaska and Herman 1986; Blaschke and Baeumler 1989). Maser et al. (1978a) postulated that edge-dwelling mammals may inoculate nonforested areas with spores of mycorrhizal fungi and thereby assist establishment of tree species. Mammals that rarely venture outside established forests, on the other hand, enhance the maintenance of high levels of inoculation in tree roots within these stands. It is certain that the role of small mammals has been underestimated despite the fact that quantitative data on their importance in different phases of the forest cycle await further investigation. There may be more ways to control small rodents in forest ecosystems than to poison them after improper management.

4 Interactions Between Burrowing Rodents and Other Animals in the Community

Since open grasslands, semideserts, and deserts do not offer protection against pedators or adverse weather conditions, every hole in the ground is a valuable resource for a variety of animals that are unable to dig burrows themselves Figs. 9–11. Others may prefer existing burrows instead of digging their own. In the southern African savanna holes dug by gerbils (*Tatera brantsii*) and springhares (*Pedetes capensis*) are used by three species of spiders (Heidger 1988), besides other invertebrates (Korn, unpubl.), and by a variety of small vertebrates like toads, snakes, and lizards (Korn and Heidger, unpubl.). The relationship between

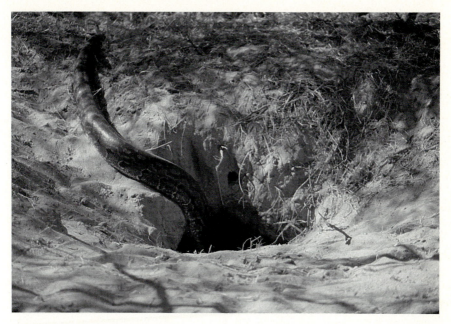

Fig. 9. A python (*Python sabae*) entering an old warthog hole after disturbance. A similar behaviour can be observed with smaller snakes that withdraw into rodent burrows; Nylsvley Nature Reserve, May 1984

Fig. 10. A lizard (*Ichnotropis capensis*) in front of a gerbil hole. The animal withdraws into the small mammal burrow when disturbed; Nylsvley Nature Reserve. August 1986

Fig. 11. A spider (*Agelena ocelata*) uses old gerbil holes for protection and to construct their web; Nylsvley Nature Reserve, August 1986

these animals and their hosts is unknown but snakes and especially lizards were more often encountered in gerbil colonies than outside. When they were disturbed they always disappeared in a small mammal hole, suggesting a certain dependence on them. The snakes probably also focus on the rodent colonies as an area of high food concentration. A similar relationship was also proposed by Osborn and Allen (1949) for rattlesnakes living in burrows of prairie dogs. Gentry and Smith (1968) found that all snakes in *Peromyscus polionotus* burrows, except the common hognose and eastern diamondback rattlesnake, were in their shedding phase.

Accounts of vertebrate and invertebrate associates of mammal burrows are given by De Graaf and Nel (1965) for the African rodent *Parotomys brantsii*, by Hickman (1984) for the mole *Scalopus aquaticus* in Florida, by Voronov (1968) for the European mole *Talpa europaea*, by Campbell and Clark (1981) for prairie dogs *Cynomys leucurus* and *C. ludovicianus*, by Gentry and Smith (1968) for the oldfield mouse *Peromyscus polionotus*, by Chandna and Pasahan (1981) for the Indian gerbil *Tatera indica*, by Bruch (1937) for the South American tuco-tuco *Ctenomys talarum*, and by Hubbell and Goff (1939), Vaughan (1961), Funderburg and Lee (1968), and Hickman (1977) for pocket gophers in North America. The list is far from complete, but the articles show the range of systematic groups encountered in mammal burrows. The invertebrate group contains insects from different orders, which may be parasitic or commensal; arachnids, including spiders, mites, harvestman, and scorpions; oligochaetes, nematodes and molluscs. The vertebrates living in mammal burrows are amphibians, reptiles, and other small mammals. Even some birds use animal holes as underground breeding sites, e.g., the blue swallow (*Hirundo atrocaerulea*) in southern Africa (McLachlan and Liversidge 1978) or the burrowing owl (*Athene cunicularia*) of western North America, which is normally closely associated with prairie dogs (Robbins et al. 1983).

Agnev et al. (1986) found significant differences between the small mammal and bird faunas inside and outside of prairie dog colonies. Furthermore, prairie voles (*Microtus ochrogaster*) were not found within dog towns and the horned lark (*Eremphila alpestris*) occurred almost exclusively within the disturbed, low vegetation sites. Campbell and Clark (1981) discussed the relationships between prairie dogs and associated fauna which they encountered in their study area. They found some predator-prey relationships between prairie dogs and birds, but also between prairie dogs and other mammals. Badgers (*Taxidea taxus*) prey heavily on prairie dogs but they are less specialized than the black-footed ferret (*Mustela nigripes*), which is almost totally dependent on them.

Turner et al. (1973) reported on the relationship between gophers to mountain voles (*Microtus montanus*) and deer mice (*Peromyscus maniculatus*). Reduction in plant litter, which is caused by gopher disturbance, makes the habitat more suitable for deer mice but less suitable for voles. Mielke (1977) presented a possible interaction with mutual benefits between large and small mammals in the prairie. He stated that the activity of the bison and the gopher complemented each other. The bison grazed and trampled the dense prairie vegetation, accelerating forb growth, on which the gophers thrive. The gopher, in turn, worked the soil, thus increasing soil fertility and stimulating vegetation growth, thus providing food for the bison.

In patches created by prairie dogs, plant biomass has a greater life-to-death ratio, a higher crude protein concentration, and a greater digestibility than biomass from uncolonized prairie (Coppock et al. 1983a). These characteristics result in improved nutrition per unit food consumed in the colony (Coppock et al. 1983b; Krueger 1986), and they may explain feeding site selection for prairie dog colonies by bison (Whicker and Detling 1988).

An interesting example for the dependence of grasshoppers on gopher mounds was shown by Huntly and Inouye (1988). The tunneling and mound-building of gophers produce a heterogeneous mosaic of physical conditions and vegetation that is beneficial to grasshoppers. Most grasshoppers oviposit in open soil, where the probability of survival of eggs and nymphs is greatest. Since grasshopper eggs were primarily found in gopher mounds they appear to increase the successful recruitment of grasshoppers and to result in higher densities than might otherwise occur. Additionally, the small-scale mosaic of bare areas, i.e., the low and high density vegetation that occurs where gophers are active, is favorable for the poikilotherm insects for thermoregulation and the efficient exploitation of the vegetation.

Vaughan (1961) recorded 22 species of vertebrates using geomyid burrows for shelter, protection, or as access routes for feeding. He noted that geomyid rodents and tiger salamanders (*Amblysoma tigrina*) appear to be tolerant to each other. In fact, the existence of the salamander in certain semiarid habitats may depend on the presence of the burrows.

The list of examples is far from complete but it shows clearly that the association of other animals with mammal burrows is a worldwide phenomenon. The only limit seems to be the distribution of the systematic groups that are dependent or favored by the presence and activity of burrowing mammals. Many of these animal species (vertebrates and invertebrates) may depend directly on the presence of mammal burrows or soil heaps. Others may not depend exclusively on them but are favored by their presence. It is certain that the animal disturbances change in time and space. The dependent animals follow and are therefore a part of the mosaic-cycle system.

5 Population Processes of Small Mammals

It is generally accepted knowledge that colonial rodents occur in high densities within colonies but are rarely encountered outside. The movements of such high density patches across the savanna, prairie, or steppe are very obvious because they are accompanied by a severe alteration of vegetation heights, vegetation density, and plant composition. These observed movements of high density patches are the basis for a new hypothesis to explain some cyclic population patterns in small, noncolonial rodents whose high density patches are not as obvious as in prairie dogs or gerbils. Up to now small mammal population studies are severely hampered by the fact that grids or trap lines are small, extending at the most over a few hectares. In high density populations, like those of *Microtus* species, it may even be a fraction of a hectar (Korn and Taitt 1987) and in many studies the home range of small rodents may be as large or larger than a traditional grid (for review, see Korn 1986).

Everything occurring outside the grid remains undetected. The population is by definition an assemblage of the animals living in the grid. Until recently, it was almost impossible to distinguish between animals born in the grid and those that immigrated. The same holds true for death and emigration. Animals that moved from the grid are traditionally treated as dead and animals moving into the grid as born on the site. If we now assume that the entire population, or a great part of it, moves from the grid when resources are depleted, and the area needs several years to recover until the population moves into the grid again, we obtain very good population cycles in the grid without a change in numbers of animals in the entire population. The same cycle can be observed when the grid is located in a suboptimal habitat that can only carry a high density population for a short time until it is depleted. The population, or part of it, may withdraw into an optimal habitat and invade the suboptimal habitat only after resources have recovered. In a grid that is placed in such an optimal habitat populations will fluctuate less than in a suboptimal habitat. This situation was encountered with small European woodland rodents near the Department of Biology of the Philipps University in Marburg, FRG (Korn, pers. observ.). Further supporting arguments come from the literature. Charnov and Finerty (1980) wrote that it has become increasingly clear that the general pattern for cyclic microtines is to persist on restricted "islands" of favorable habitats during years of population lows, and then to disperse into less favorable habitats as numbers begin to increase. It is possible to assume that these restricted "islands" mentioned above will not be the same over time since they represent a certain point in the plant succession which is favorable for the animals. Again we find a mosaic of habitats of different quality that presumably change their location over time. The grid is stationary but the small mammal population may move over it like a wave. An example for such a system has been shown by Andreev (1988) for willow grouse (*Lagopus lagopus*) in Sibiria. In normal years birds occupy only 3 to 4% of the whole area, which consists of the preferred habitat, while during the population peak, the less preferred parts of the tundra are also occupied. For 2 to 3 years birds use vast areas with a low density of suitable food to support reproduction. When the resouces are exhausted they take a long time to recover. Meanwhile the grouse population is restricted to patches where willows of varying quality are abundant. Only a few of them (less than 1% of the total area) may permanently support a high density of willow grouse. The principal breeding areas of the birds were not the same every year with a wave of density moving northward. I suppose that the famous lemming migrations and their pronounced population cycles correspond to the same general system as the willow grouse. Small mammals are much more difficult to observe than birds, but the displacement of entire small rodent populations can be extensive as reported for gerbils (*Gerbilus* sp.) and ground squirrels (*Xerus inauris*) in the Namib desert (Cox 1987). Pitelka (1973) pointed out that events in neighboring lemming populations, not in-phase with those of Barrow (Alaska) but promoting dispersal to it, can explain some population processes. He suspects that the Peard Bay area (95 km away) could have been the source of lemmings which showed up near Barrow in late winter of 1971. The proposed mechanism of a rodent population moving across the grid or trap lines and creating a population cycle is certainly not a conclusive explanation for all cyclic population processes or population outbreaks in small mammals. It is rather an explanation of an artifact encountered due to poor methodology. Some results may have to be reconsidered in this respect. To study the fate of individuals and entire grid populations we need huge study areas that cover several large vegetation mosaics or better methods to track small mammals than we have now.

6 Conclusion

As we have seen above, the relationship between small mammals and the mosaic-cycle concept of ecosystems is manifold. Mammals can be the driving force of the cycle, e.g., the burrowing mammals in low vegetation ecosystems, or may passively follow changes in the environment caused by storms, fire, plant succession, or forestry practice in forest ecosystems. Nevertheless, it should be kept in mind that the latter conclusion may change when more information on the importance of small mammals in natural forest ecosystems becomes available. They may, as in low vegetation ecosystems, play a major role in the maintenance of soil fertility and species diversity of plants and animals. In grasslands and herb lands small mammals are capable of destroying entire plants because the low growth form makes all parts accessible to them and vulnerable to destruction. Thus, they open up sites for competitively, inferior plants and create sites of interest for animals. The large size and life span may put trees and shrubs beyond the influence of small mammals except when they are small. Therefore, in forest communities the influence of small mammals seems to concentrate on regeneration. The regenerative phase of the forest cycle with its seedlings and small trees is very similar to the low vegetation ecosystems and small mammals play an important role until the trees have reached a certain size.

Fieldwork in intact systems is desperately needed to fill the incredible gap in our knowledge of the importance of small mammals in ecosystems and their influence on natural cycles. This is not only necessary because of scientific interest but in order to develop a new integrated concept for forest and agricultural practice which is needed to increase life quality for humans, to enhance species and population survival of plants and animals, and to secure natural resources for the future.

Only intact systems will provide all these benefits. We should learn from nature to see how she recovers from natural catastrophes and use that knowledge to develop proper management practices that do not irreparably damage the system. Only that knowledge, for which ecologists with a sound education are needed, can secure the indefinite use of natural resources. Technocrats, with their very narrow view of a system, backed by economists that can only think in categories of short-term profit, have done enough damage in the past and keep doing it in the present. With our present environmental problems it is obvious that expensive technical solutions of problems, created by improper management, fail when the system is not understood. The reality is that for expensive technical "solutions", which often aggravate the problem there are always enough funds, but for comparatively inexpensive, thoughtful investigations on proper management practices there are none. Often it is better to do nothing than to practice poor management. We are so used to quick technical solutions that it is difficult to convince technocrats not to interfere when it is not necessary. It is often unnecessary and we can sit back and wait, which is also much cheaper, when we understand the system.

We all, not only researchers and environmentalists, need intact natural systems as a baseline to determine what we have done wrong regarding managed systems. If every single square kilometer of this earth is clear-cut, overgrazed, or turned over to improper agriculture, what comparison remains if our managed systems fail and deteriorate? All humans depend on the use of natural resources but their proper, integrated management is needed to provide a sustainable yield.

References

Abaturov BC (1964) The influence of *Lagurus lagurus* on the soil and vegetable cover in the dry steppe of Kazakhstan. Biull Moskov Obshch Ispyt Prirod Otd Biol 69:24–35

Abaturov BD (1972) The role of burrowing animals in the transport of mineral substances in the soil. Pedobiologia 12:261–266

Agnev W, Uresk DW, Hansen RM (1986) Flora and fauna associated with prairie dog colonies and adjacent ungrazed mixed-grass prairie in western South Dakota. J Range Manage 39:135–139

Ahlgren CE (1966) Small mammals and reforestation following prescribed burning. J For 64:614–618

Aikman JM, Thorne RF (1956) The Caylor Prairie: an ecologic and taxonomic study of a northern Iowa prairie. Proc Iowa Acad Sci 63:177–200

Andreev A (1988) The ten year cycle of the willow grouse of Lower Kolyma. Oecologia 76:261–267

Atkeson TD, Johnson AS (1979) Succession of small mammals on pine plantations in the Georgia Piedmont. Am Midl Nat 101:385–392

Baeumler W (1986) Truffeln, Mäuse und Testosteron. Naturwiss Rundsch 39:396–397

Baeumler W, Hohenadl W (1980) Ueber den Einfluss alpiner Kleinsaeuger auf die Verjuengung in einem Bergmischwald der Chiemgauer Alpen. Forstwiss Zentralbl 99:207–221

Berman DI, Kuz'min IF, Tichomirova LG (1966) Burrowing activities of animals in the SE Tuva plain pastures. In: Organismy i prirodnaja sreda, pp 60–75

Blaschke H, Baeumler W (1986) Ueber die Rolle der Biogeozoenose im Wurzelbereich von Waldbaeumen. Forstwiss Zentralbl 105:122–130

Blaschke H, Baeumler W (1989) Mycophagy and spore dispersal by small mammals in Bavarian forests. For Ecol Manage 26:237–245

Bonham CD, Lerwick A (1976) Vegetation changes induced by prairie dogs in shortgrass range. J Range Manage 29:221–225

Borrecco JE (1976) Controlling damage by forest rodents and logomorphs through habitat manipulation. In: Siebe CC (ed) Proc 7th Vertebr Pest Conf March 9–11, 1976, Monterrey, California, pp 203–210

Borrecco JE, Anderson RJ (1980) Mountain beaver problems in, the forests of California, Oregon and Washington. In: Clark JP (ed) Proc 9th Vertebr Pest Conf, March 4–6, 1980, Fresno, California, pp 135–142

Bruch C (1937) Notas etologicas acerca del "tuco-tuco" (Ctenomys talarum talarum O. Thomas) y nomina de artropodos que viven en sus habitaculos. Notas Mus La Plata 2, Zool 6:81–86

Buechner HK (1942) Interrelationship between the pocket gopher and land use. J Mamm 23:346–348

Burggraaf van Nierop YA, van der Meijden E (1984) The influence of rabbit scrapes on dune vegetation. Biol Conserv 30:133–146

Campbell TM, Clark TW (1981) Colony characteristics and vertebrate associates of white-tailed and black-tailed prairie dogs in Wyoming. Am Midl Nat 105:269–275

Capp JC (1976) Increasing pocket gopher problems in reforestation. In: Siebe CC (ed) Proc 7th Vertebr Pest Conf March 9–11, 1976, Monterrey, California, pp 221–228

Carlson DC, White EM (1987) Effects of prairie dogs on mound soils. Soil Sci Soc Am J 51:389–393

Chandna SS, Pasahan SC (1981) Field observations on the ecology of the Indian gerbil Tatera indica (Hardwicke) in Hissar (Haryana). Ann Arid Zone 20:62–70

Charnov EL, Finerty JP (1980) Vole population cycles: a case for kin-selection? Oecologia 45:1–2

Chesemore DL (1969) Den ecology of the Arctic fox in northern Alaska. Can J Zool 47:121–129

Contreras JR, Maceiras AJ (1970) Relaciones entre tucu-tucos y los procesos del suelo en la region semiarida del sudoeste bonaerense. Agro 12 (17):1–26

Coppock DL, Detling JK, Ellis JE, Dyer MI (1983a) Plant-herbivore interactions in a North American mixed-grass prairie I. Effects of black-tailed prairie dogs on intraseasonal aboveground plant biomass and nutrient dynamics and plant species diversity. Oecologia 56:1–9

Coppock DL, Ellis JE, Detling JK, Dyer MI (1983b) Plant-herbivore interactions in a North American mixed-grass prairie II. Responses of bison to modification of vegetation by prairie dogs. Oecologia 56:10–15

Cox GW (1987) The origin of vegetation circles on stoney soil of the Namib desert near Gobabeb, South West Africa, Namibia. J Arid Envir 13:237–243

Crouch GL (1982) Pocket gophers and reforestation on western forests. J For 80:662–665

Crouch GL (1986) Pocket gopher damage to conifers in western forests: a historical and current perspective on the problems and its control. In: Salmon TP (ed) Proc 12th Vertebr Pest Conf San Diego, California, March 4–6, 1986, Univ California, Davis, pp 196–198

Dallsted KJ, Sather-Blair S, Worcester BK, Klukas R (1981) Application of remote sensing to prairie dog management. J Range Manage 34:218–223

DeGraaf G, Nel JAJ (1965) On the tunnel system of Brandt's karroo rat, Parotomys brantsii in the Kalahari Gemsbok National Park. Koedoe 8:136–139

Downhower JF, Hall ER (1966) The pocket gopher in Kansas, Univ Kansas Mus Nat Hist Misc Publ 44:1–32

Dusen P (1903) Die Pflanzenvereine der Magellanslander nebst einem Beitrag zur Ökologie der magellanischen Vegetation. Svenska Expeditionen till Magellanslanderna III (10):382–384

Ellison L (1946) The pocket gopher in relation to soil erosion on mountain range. Ecology 27:101–114

Ellison L, Aldous CM (1952) Influence of pocket gophers on vegetation of subalpine grassland in central Utah. Ecology 33:177–186

Fogel R, Trappe JM (1978) Fungus consumption (mycophagy) by small mammals. Northwest Sci 52:1–31

Formozov AN (1928) Mammalia in steppe biocenose. Ecology 9:449–460

Formozov AN (1978) Mammalia in steppe biocenose. Byulleten Mosk Obshch Ispyt Pris Otd Biol. 83:150–156

Foster MA, Stubbendieck J (1980) Effects of the plains pocket gopher (Geomys bursarius) on rangelands. J Range Manage 33:74–80

Funderburg JB, Lee DS (1968) The amphibian and reptile fauna of pocket gopher (*Geomys*) mounds in central Florida. J Herpetol 1:99–100

Gashwiler JS (1970) Plant and mammal changes on a clearcut in west-central Oregon. Ecology 51:1018–1026

Gentry JB, Smith MH (1968) Food habits and burrow associates of *Peromyscus polionotus*. J Mammal 49:562–565

Grant WE, McBrayer JF (1981) Effects of mound formation by pocket gophers (*Geomys bursarius*) on old-field ecosystems. Pedobiologia 22:21–28

Grant WE, French NR, Folse LJ (1980) Effects of pocket gopher mounds on plant production in shortgrass prairie ecosystems. Southwestern Nat 25:215–224

Greene RA, Murphey GH (1932) The influence of two burrowing rodents, *Dipodomys spectabilis spectabilis* (kangaroo rat) and *Neotoma albigula albigula* (pack rat) on desert soils in Arizona. Ecology 13:73–80

Greene RA, Reynard C (1932) The influence of two burrowing rodents, *Dipodomys spectabilis spectabilis* (kangaroo rat) and *Neotoma albigula albigula* (pack rat) on desert soils in Arizona. II. Physical effects. Ecology 13:359–363

Grinnell J (1923) The burrowing rodents of California as agents in soil formation. J Mamm 4:137–149

Grulich I (1959) Wuehltaetigkeit des Maulwurfs in der Tschechoslowakei. Acta Acad Sci Cechoslov Basic Brun 31:157–216

Gutterman Y, Herr N (1981) Influence of porcupine (*Hystrix indica*) activity on the slopes of the northern Negev Mountains — germination and vegetation renewal in different geomorphological types and slope directions. Oecologia 51:332–334

Hansen RM, Morris MJ (1968) Movements of rocks by northern pocket gophers. J Mamm 49:391–399

Heidger D (1988) Ecology of spiders inhabiting abandoned mammal burrows in South African savanna. Oecologia 76:303–306

Herrchen S (1989) Ökologische Untersuchungen an Kleinsäugern auf Windwurfflächen des frühen Sukzessionsstadiums im Nationalpark Bayrischer Wald. Populationsdynamik, Mobilität und Habitatpräferenz: Vergleich mit Naturwald der Optimalphase. Diplomarbeit, Fachbereich Biologie, Philipps-Universitaet Marburg, FRG

Hickman GC (1977) Geomyid interaction in burrow systems. Texas J Sci 29:235–244

Hickman GC (1984) An excavated burrow of *Scalopus aquaticus* from Florida, with comments on Nearctic talpoid/geomyid burrow structure. Saugetierkd Mitt 31:243–249

Hobbs J, Mooney HA (1985) Community and population dynamics of serpentine grassland annuals in relation to gopher disturbance. Oecologia 67:342–351

Hooven EF (1971) Pocket gopher damage on ponderosa pine plantations in southwestern Oregon. J Wildlife Manage 35:346–353

Hsia W, Chong W (1966) The succession and interactions of rodents and plant communities of abandoned fields in desert steppe at Chagan-Aabar, Inner Mongolia. Acta Zool Sinica 18:199–208

Hubbell TH, Goff CC (1939) Florida pocket gopher burrows and their arthropod inhabitants. Q J Fla Acad Sci 4:127–166

Huntly N, Inouye R (1988) Pocket gophers in ecosystems: patterns and mechanisms. BioScience 38:786–793

Ingles LG (1952) The ecology of the mountain pocket gopher, *Thomomys monticola*. Ecology 33:87–95

Inouye RS, Huntly NJ, Tilman D, Tester JR (1987) Pocket gophers (*Geomys bursarius*), vegetation, and soil nitrogen along a successional sere in east central Minnesota. Oecologia 72:178–184

Jensen TS (1984) Habitat distribution, home range and movements of rodents in mature forest and reforestations. Acta Zool Fenn 171:305–307

Jettmar HM (1926) Die Bauten einiger Transbaikalischer Säugetiere in schematischer Darstellung. Z Saugetierk 1:14–34

Kalisz P, Stone P (1984) Soil mixing by scrab beetles and pocket gophers in north-central Florida. Soil Sci Soc Am J 48:169–172

King JA (1955) Social behavior, social organization, and population dynamics in a black-tailed prairie dog town in the Black Hills of South Dakota. Contrib Lab Vertebr Biol. Univ Michigan, 67:1–123

Kiryushchenko SP (1978) Effects of the digging activity of the lemming *Discrostonyx torquatus* Pall. on the vegetational cover of the Arctic tundras of Wrangel Island. Byulleten Mosk Obshch Ispyt Pris Otd Biol 83:28–35

Klatt LE, Hein D (1978) Vegetative differences among active and abundant towns of black-tailed prairie dogs (*Cynomys ludovicianus*). J Range Manage 31:315–317

Knapp R (1959) Untersuchungen ueber den Einfluss von Tieren auf die Vegetation. 1. Rasen-Gesellschaften und Talpa europaea. Angew Bot 33:177–189

Koide RT, Huenneke LF, Mooney HA (1987) Gopher mound soil reduces growth and affects ion uptake of two annual grassland species. Oecologia 72:284–290

Korn H (1986) Changes in home range size during growth and maturation of the wood mouse (*Apodemus sylvaticus*) and the bank vole (*Clethrionomys glareolus*). Oecologia 68:623–628

Korn H, Korn U (1989) The effect of gerbils (*Tatera brantsii*) on primary production and plant species composition in a southern African savanna. Oecologia 79:271–278

Korn H, Taitt MJ (1987) Initiation of early breeding in a population of *Microtus townsendii* (Rodentia) with the secondary plant compound 6-MBOA. Oecologia 71:593–596

Kotter MM, Farentinos RC (1984) Tassel-eared squirrels as spore dispersal agents of hypogeous mycorrhizal fungi. J Mamm 65:684–687

Kozhemyakin VV (1978) Change in the vegetation cover around rodent burrows in the desert belt of Mongolia. Problemy Osvonija Pustyn 1:80–85

Krueger K (1986) Feeding relationships among bison, pronghorn, and prairie dogs: an experimental analysis. Ecology 67:760–770

Lavrenko EM (1952) Microkomplexe und Mikromosaik der Pflanzendecke der Steppe als Folge der Lebenstaetigkeit der Tiere unf Pflanzen. Arb Bot Inst Acad Wiss Leningrad Ser III Geobotanik 8:40–70

Laycock WA (1958) The initial pattern of revegetation of pocket gopher mounds. Ecology 39:346–351

Laycock WA, Richardson BZ (1975) Long-term effects of pocket gopher control on vegetation and soils of a subalpine grassland. J Range Manage 28:458–462

Leverich WJ (1983) Pocket gophers, tanks, and plant community composition. Southwestern Nat 28:378

Löffler E, Margules C (1980) Wombats detected from space. Remote Sens Environ 9:47–56

Mallory FF, Boots BN (1983) Spatial distribution of lemming mats in the Canadian High Arctic. Can J Zool 61:99–107

Martell AM (1983) Changes in small mammal communities after logging in north-central Ontario Can J Zool 61:970–980

Maser C, Tappe JM, Nussbaum RA (1978a) Fungal-small mammal interrelationships with emphasis on Oregon conifer forests. Ecology 59:799–809

Maser C, Tappe JM, Ure DC (1978b) Implications of small mammal mycophagy to the management of western conifer forsts. Trans Am Wildl Nat Res Conf 43:78–88

McDonough WT (1974) Revegetation of gopher mounds on aspen range in Utah. Great Basin Nat 34:267–274

McKeever S (1961) Relative populations of small mammals in forest types of northeastern California. Ecology 42:399–402

McKendrick JD, Batzli GO, Everett KR, Swanson JC (1980) Some effects of mammalian herbivores and fertilization on tundra soils and vegetation. Arct Alp Res 12:565–578

McLachlan GR, Liversidge R (1978) Roberts birds of South Africa, 4th ed. Voelcker Bird Book Fund, Cape Town, 659 pp

Merriam CJ (1902) The prairie dog of the great plains. USDA yearbook 1902:257–270

Mielke HW (1977) Mound building by pocket gophers (Geomyidae): The impact on soil and vegetation in North America. J Biogeogr 4:171–180

Miller MA (1957) Burrows of the Sacramento Valley pocket gopher in flood-irrigated alfalfa fields. Hilgardia 26:431–452

Mohr CD, Mohr WP (1936) Abundance and digging rate of pocket gophers, *Geomys bursarius*. Ecology 17:325–327

Moroka N, Beck RF, Pieper RD (1982) Impact of burrowing activity of the bannertail kangaroo rat on southern New Mexico desert rangelands. J Range Manage 35:707–710

Naumov NP, Lobachev VS (1975) Ecology of desert rodents in the USSR. In: Prakash I, Gosh PK (eds) Rodents in desert environments. Junk, The Hague, pp 465–598

Nixon CM, Worley DM, McClain MW (1968) Food habits of squirrels in southeast Ohio, J Wildl Manage 32:294–305

Osborn B, Allan P (1949) Vegetation of an abandoned prairie dog town in tall grass prairie. Ecology 30:322–332

Ovaska K, Herman TB (1986) Fungal consumption by six species of small mammals in Nova Scotia. J Mamm 67:208–211

Pitelka FA (1973) Cyclic pattern in lemming populations near Barrow, Alaska. In: Britton ME (ed) Alaskan arctic tundra. Arct Inst North America, Techn Pap 25:199–215

Platt WJ (1975) The colonization and formation of equilibrium plant species associations on badger disturbances in a tall-grass prairie. Ecol Monogr 45:285–305

Ramirez P, Hornocker M (1981) Small mammal populations in different-aged clearcuts in north-western Montana. J Mamm 62:400–403

Reichman OJ, Jarvis JUM (1989) The influence of three species of fossorial mole-rats (*Bathyergidae*) on vegetation. J Mamm 70:763–771

Reichman OJ, Smith SC (1985) Impact of pocket gopher burrows on overlying vegetation. J Mamm 66:720–725

Reig OA (1970) Ecological notes on the fossorial octodont rodent *Spalacopus cyanus* (Molina). J Mamm 51:592–602

Remmert H (1985) Was geschieht im Klimax-Stadium? Naturwissenschaften 72:505–512

Remmert H (1988) Gleichgewicht durch Katastrophen. Aus Forschung und Medizin 3 (1):7–17

Richens VB (1966) Notes on the digging activity of a northern pocket gopher. J Mamm 47:531–533

Robbins CS, Brunn B, Zim HB (1983) Birds of North America, 2nd ed. Golden Press, New York, 360 pp

Rodin LE (1961) Dynamics of desert vegetation as exemplified in western Turkmenistan. ANSSR, Bot Inst Leningrad, 227 pp

Schaal BA, Leverich WJ (1982) Survivorship patterns in an annual plant community. Oecologia 54:149–151

Schauer J (1987a) The effect of the vole *Microtus brandti* on the Mongolian steppes. Folia Zool 36:203–214

Schauer J (1987b) Remarks on the construction of burrows of *Ellobius talpinus, Myospalax aspalax* and *Ochotona daurica* in Mongolia and their effects on soil. Folia Zool 36:319–326

Senzota RBM (1984) The habitat, abundance and burrowing habits of the gerbil *Tatera robusta*, in the Serengeti National Park, Tanzania. Mammalia 48:185–195

Sharma VN, Joshi MC (1975) Soil excavated by desert gerbils *Meriones huriane* (Jerdon) in the Shekhawati region of Rajastan desert. Ann Arid Zone 14:268–273

Skoczen S, Szot T, Dabrowska L (1983) The effect of the activity of the mole (*Talpa europaea*) on the floristic composition of the grassland swards of southern Poland. Acta Agrar Silvestria (Ser) Agrar 22:101–120

Smith CF, Aldous SE (1947) The influence of mammals and birds in retarding artificial and natural reseeding of conifer forests in the United States. J For 45:361–369

Smith RE (1967) Natural history of the prairie dog in Kansas. Univ Kansas Mus Natur Hist State Biol Surv Misc Publ 49:1–39

Spencer SR, Cameron GN, Eshelman BD, Cooper LC, Williams LR (1985) Influence of pocket gopher mounds on a Texas coastal prairie. Oecologia 66:111–115

Stogova LL (1986) Effect of the mole voles on the structure and productivity of mountain pastures based on the Assy Tract Zailiiskii Ala Tau Kazakh SSR, USSR. Izv Acad Nauk Kaz SSR Ser Biol O (5):25–27

Sullivan TP (1979) Repopulation of clear-cut habitat and conifer seed predation by deer mice. J Wildl Manage 43:861–871

Taylor WP (1924) Damage to range grasses by the zuni prairie dog. US Dept Agric Bull 1227:1–15

Taylor WP (1935) Some animal relations to soils. Ecology 16:127–136

Teipner CL, Garton EO, Nelson L (1983) Pocket gophers in forest ecosystems. Department of Agriculture Forest Service. Intermountain Forest and Range Experimental Station. Ogden, UT 84401. Gen Techn Rep INT-154:53

Tevis L Jr (1956a) Pocket gophers and seedlings of red fir. Ecology 37:379–381

Tevis L Jr (1956b) Effect of a slash burn on forest mice. J Wild Manage 20:405–409

Thompson MP (1972) Feeding and burrowing effects of woodchucks (*Marmota monax*) on old field vegetation. Diss Abstr Int 32 B (9):5546–5547

Tilman D (1983) Plant succession and gopher disturbance along an experimental gradient. Oecologia 60:285–292

Turner GT, Hansen RM, Reid HP, Tietjen HP, Ward AL (1973) Pocket gophers and Colorado mountain rangeland. Colorado State Univ Exp Sta, Fort Collins, Bull 554:90

Vaughan TA (1961) Vertebrates inhabiting pocket gopher burrows in Colorado. J Mamm 42:171–174

Voronov NP (1968) Ueber die Wuehltaetigkeit des Maulwurfs (*Talpa europaea* L.). Pedobiologia 8:97–122

Walter H, Breckle H (1983) Oecologie der Erde Bd. 1. Oecologische Grundlagen in globaler Sicht. Fischer, Stuttgart

Walter H, Breckle H (1986) Oecologie der Erde Bd. 3. Spezielle Oekologie der gemaessigten und arktischen Zonen Euro-Nordasiens. Fischer, Stuttgart

Weiner J, Gorecki A (1982) Small mammals and their habitat in the arid steppe of central eastern Mongolia. Polish Ecol Stud 8:7–21

Weiner J, Gorecki A, Zielinski J (1982) The effect of rodens on the rate of matter and energy cycling in ecosystems of arid steppe of central eastern Mongolia. Polish Ecol Stud 8:69–86

Werner DJ (1977) Vegetationsveränderungen in der argentinischen Puna unter dem Einfluss von Bodenwühlern der Gattung *Ctenomys* Blainville. In: Tuxen R (ed) Vegetation und fauna. Vaduz 1977, pp 433–449

Whicker AD, Detling JK (1988) Ecological consequences of prairie dog disturbances. BioScience 38:778–785

Wiens JA (1985) Vertebrate responses to environmental patchiness in arid and semiarid ecosystems. In: Picket STA, White PS (eds) The ecology of natural disturbances and patch dynamics. Academic Press, Orlando, pp 169–193

Williams LR, Cameron GN (1986) Effects of removal of pocket gophers on a Texas coastal prairie. Am Midl Nat 115:216–224

Williams LR, Cameron GN, Spencer SR, Eshelman BD, Gregory MJ (1986) Experimental analysis of the effects of the pocket gopher mounds on Texas coastal prairie. J Mamm 67:672–679

Xia W, Zhong W (1978) On the relationship between the dens of the ground squirrel and the plant community. Acta Zool Sin 24:335–343

Zaletayev VS (1976) Life in deserts. (Geographical-Biocenological and Ecological Problems. Mysl, Moscow, 271 pp

Zhuchkova VK, Utekhin VD (1975) Effects of burrowing activity of *Spalax microphthalmus* Gueld. on the vegetation of forest-steppe biogeocenoses. Byulleten Mosk Obschk Ispyt Prir Otd Biol 80:134–145

Zielinski J (1982) The effect of brandt vole (*Microtus brandti* Raddi, 1861) colonies upon vegetation in the Caragana steppe on central eastern Mongolia. Polish Ecol Stud 8:41–56

Zonov GV, Naumenko AA, Smagulov TA (1983) Soil mantle differentiation in the Turgay Plateau dry steppes. Pocvovedenie O (12):17–24

Phytoplankton:
Directional Succession and Forced Cycles

U. SOMMER

1 Introduction

Plankton communities undergo conspicuous temporal change in species composition. Such change occurs over long-term scales in water bodies undergoing environmental change (e.g., eutrophication) and within individual years. The latter has been termed "seasonal succession" by plankton ecologists. Despite its cyclic character, plankton succession has more similarities with succession than with seasonal apsectation in terrestrial vegetation: Numerous generations are involved; the abundance of individual populations and, thus, the species composition of the community undergo drastic change; community composition passes several quite distinct stages; rescaled to the generation time of the organism and the duration time of stages several months of plankton succession are analogous to several centuries of forest succession.

Initially, the seasonal change in plankton communities was viewed as the composite result of independent responses of the individual populations to external environmental conditions, primarily light and temperature. Later, biotic interactions became more appreciated, with the emphasis placed on exploitative competition (Tilman 1982) and on alga-grazer interactions (Porter 1977). Because of the increasing recognition of biotic interactions, it has become customary to distinguish successional from nonsuccessional changes in species composition (Reynolds 1980). Successions are those changes in species composition which result from biotic interactions under externally unperturbed conditions, i.e., under constant or increasing physical stability of the water column. Selection under such conditions increasingly favors species which are good competitors for scarce resources and/or which are resistant against grazing. Physical perturbation (loss of water column stability, increase of mixing depth) leads to a partial or complete release from those selective pressures and to nonsuccessional changes in species composition. Such changes are termed "reversion" if the release from competition and predation pressure is short and a new succession is initiated immediately afterwards. Such reversions are caused by surface cooling and increased wind mixing in the middle of the vegetation period and have a typical time scale of 5 to 15 days in most of the temperate region. Nonsuccessional changes which occur more continuously on longer time scales

Max-Planck-Institut für Limnologie, Postfach 165, 2320 Plön, Germany

H. Remmert (Ed.)
Ecological Studies Vol. 85
© Springer-Verlag Berlin Heidelberg 1991

are termed "shifts". A typical example is the autumn situation in lakes, where decrease of day length, surface irradiance and water temperature and increase of mixing depth override the selective importance of biotic interactions. Thus, succession is viewed as being unidirectional, while the cyclic character of phytoplankton seasonality is viewed as being externally forced.

Traditional issues of succession theory ("directionality", "maturity", "climax", etc.) carry a heavy load of the Clementsian myth of the "superorganism" and need to be redefined in order to become operational for plankton research. The question of "directionality" becomes "predictability", i.e., whether there are repeatable patterns under similar conditions. This question may be asked at the level of taxonomic composition, but also at the level of more aggregated variables such as production/biomass ratios, diversity, etc. The term "maturity" has no real meaning outside the concept of the superorganism, "immature vs mature" reduces to "early vs late successional". The monoclimax vs polyclimax debate has never been an issue in plankton ecology, because it seems an idle question how late summer communities would develop in the absence of autumn cooling and decrease of light intensities. It is, however, a relevant question whether plankton succession is "convergent" if the initial species composition is dissimilar and external conditions are similar. Moreover, the question may be asked whether late successional stages are self-sustainable (as hypothesized by the traditional succession theory) or bound to collapse (as hypothesized by the mosaic-cycle theory).

2 The Directionality of Phytoplankton Succession

During the recent years evidence for the regularity of successional patterns has increased (Reynolds 1980, 1984, 1988; Sommer 1986). It was not only possible to extract a relatively small number of successional pathways from the plethora of available data, but it has also been possible to formulate hypotheses about causal mechanisms and to test them by culture experiments, field experiments, and by measurement of in-situ processes. Thus, the importance of stratification (Reynolds et al. 1983), sedimentation losses (Reynolds and Wiseman 1982; Sommer 1984), competition for resources (Tilman 1982; Sommer 1989), mortality and selective impact by grazing (Sterner 1989), and interactions between competition and grazing (Sommer 1988) are experimentally well-established components of phytoplankton succession theory. It has been shown by case studies (Blelham Tarn, Reynolds et al. 1982; Lake Constance, Sommer 1987) that the complex interactions between the different driving forces of phytoplankton species replacements can be disentangled and that a causal explanation of succession is possible.

During winter, phytoplankton growth is restricted by lack of light. Low incoming radiation, short day length, light extinction by ice and snow cover in lakes with ice cover, and deep water circulation in lakes without ice cover add up to severe lack of light. Release from this physical control by ice break or (lakes of moderate depth) by the onset of stratification (deep lakes) permits the start of seasonal succession, during which higher trophic levels (zooplankton) build up and one or several nutrients become depleted. Depending on hydrographic and chemical conditions, different successional pathways can be distinguished, but generally succession proceeds in the direction of species which resist grazing pressure (low edibility), which are good competitors for limiting resources (light and nutrients), or which have access to alternative resource pools (vertically migrating flagellates and Cyanobacteria, which can exploit nutrient reserves in deeper waters, N_2-fixing Cyanobacteria).

In a comparison of 25 different lakes (Sommer et al. 1986) two standard types of succession could be distinguished (Fig. 1). In eutrophic lakes there is a rapid development of an algal spring bloom, consisting of small flagellates and centric diatoms (low mixing depth) or of larger diatoms (high mixing depth). This algal bloom is followed by a rapid increase of zooplankton (mainly Cladocerans) which soon attain grazing rates in excess of algal growth rates. The subsequent crash of phytoplankton biomass ("clear-water phase") belongs to the most predictable events in many lakes. After a decrease of zooplankton caused by food limitation and fish predation phytoplankton builds up again under moderate grazing pressure. During that period grazing selects for size classes (large unicells or colonies and gelatinous algae are poorly edible) and competition for limiting

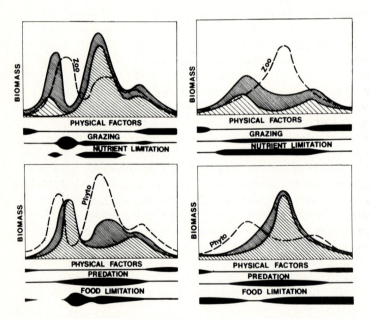

Fig. 1. The seasonal development of phytoplankton (*top panels*) and zooplankton (*bottom panels*) in an idealized stratifying eutrophic lake (*left*) and an idealized stratifying oligotrophic lake (*right*). *Top panels*: small algae (*fine hatching*); large algae (*medium hatching*); diatoms (*widely spaced hatching*); zooplankton biomass: *dashed line*. *Bottom panel*: small herbivores (*dense hatching*); large herbivores (*coarse hatching*); phytoplankton biomass (*dashed line*). The horizontal diagrams indicate the seasonal change in the importance of controlling factors (Sommer et al. 1986)

resources selects for taxa. The development in autumn is nonsuccessional and beyond the scope of this chapter. In some moderately eutrophic lakes the Cyanobacterium *Oscillatoria rubescens* is superimposed on this scheme. It forms mass developments during deep circulation and "oversummers" in dense, narrow layers in the thermocline. If the thermocline is physically unstable (e.g., Lake Constance) or too dark (e.g., Lake Zürich during the peak of eutrophication), this alga cannot persist.

In oligotrophic lakes the spring bloom of phytoplankton usually develops more slowly and large diatoms are often more prominent because of their good competitive abilities for phosphorus. Zooplankton build up much slower than in eutrophic lakes and the decline after spring is less sharp than the clear-water phase of eutrophic lakes. Due to the lack of nutrients there is no development of a summer bloom. Only the increase of mixing depth in autumn may permit a second seasonal phytoplankton peak because of import of fresh nutrients from deeper waters.

Figure 2 shows the typical phytoplankton succession in a meso- to eutrophic hard-water lake without *Oscillatoria*. The spring bloom starts with small flagellates (frequently Cryptophyceae) and unicellular centric diatoms if mixing depths are low. This is the case when the spring bloom starts with the onset of stratification (deep monomictic lakes) and the input of wind energy is low. If mixing depth is higher (start during circulation or strong winds after stratification), the average light intensity in the mixed water layer decreases and

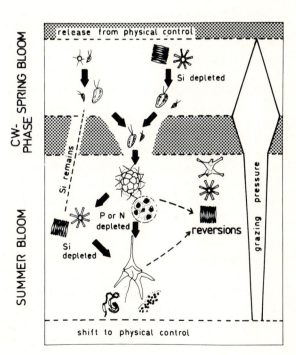

Fig. 2. Successional sequence of phytoplankton in a eutrophic lake without *Oscillatoria rubescens*. *Thick arrows*: successional changes; *dashed arrows*: reversions; *shaded* background: biomass minima

selection for larger diatoms with low light requirements occurs. This difference has consequences for the successional pathways later in the season, because the more diatoms present, the more Si will be removed and lost by sedimentation, thus becoming unavailable for later diatom growth. Later in the spring bloom, larger Cryptophycean flagellates also appear. The large diatoms are partially removed by sedimentation, while the edible small diatoms and the Crypto-phyceae are mainly removed by grazing. During the clear water phase they are still dominant, but at much lower numbers. Intensive zooplankton excretion regenerates P and N, but not Si. When grazing pressure declines large, inedible algae (e.g., colonial, gelatinous Chlorophyta) and edible flagellates increase under initially nonlimiting nutrient conditions. Depending on the chemical conditions of the lake, either N or P or both are exhausted. If there is still Si left (few diatoms during the spring bloom), diatoms will follow, because they are good competitors for P. Their development will eventually be terminated by Si-de-pletion and most of them will be lost by sedimentation. In the absence of diatoms large dinoflagellates and colonial Cyanobacteria become the dominant com-ponents of phytoplankton. Both are large, have low growth rates, are vertically motile (access to higher nutrient concentrations in deep water), and highly resistant against grazing and sinking losses. If nitrogen is depleted, N_2-fixing Cyanobacteria have an additional advantage. If deterioration of weather con-ditions leads to an increase of mixing depth, the density of algae is reduced, nutrient stress is alleviated because of imports from deeper waters, but grazing pressure remains unaffected. If such a disturbance leads to a sufficient increase of Si, diatoms will reappear, otherwise desmids and/or colonial chlorococcales will appear. Dominance by large dinoflagellates and Cyanobacteria is usually the final stage of succession, but it is unknown (and quite irrelevant) whether succession would proceed further without the shift to physical control in autumn.

3 Competitive Exclusion, Disturbance and Diversity

It has been shown repeatedly by chemostat competition experiments that only as many species can permanently coexist as there are different limiting resources (Tilman 1982; Sommer 1989). A chemostat is a flow-through culture where fresh medium is delivered at a constant rate and organisms and used medium are exported at the same rate. In such a culture eventually a steady state develops where the import of nutrients balances consumption by organisms and the reproduction of organisms balances their export. Coexistence of different species is only possible by being limited by different resources, e.g., a silicate-limited diatom could coexist with a P-limited alga (diatom or not), but two silicate-limited diatoms could not coexist with each other. The coexistence of two species based on two limiting resources requires that one species is the better competitor for one resource and the other species is the better competitor for the other resource. If many species are inversely ranked in their competitive abilities for two resources, then the same amount of species can share a gradient of the ratio of the two re-sources. However, at one particular ratio only two species would persist (Tilman

1982). They would be those two species whose "optimal ratio" of the two resources is closest to the supply ratio. The optimal ratio is defined at the transition ratio from limitation by resource 1 to limitation by resource 2. Note that the coexisting species are quite similar in their requirements for those two resources and not dissimilar, as the classic niche concept would imply.

Chemostats are an idealized model of the summer situation in the mixed surface layer of lakes. Nutrient consumption cannot rely on a big unused pool as during initial colonization in spring. It has to rely on the renewal of nutrients by zooplankton excretion, diffusive input from deeper waters, and input from the watershed, i.e., phytoplankton has to live on the income of nutrients, not on the capital. Similarly, there is a constant drain on phytoplankton populations by grazing and sinking losses, so that most of the reproduction just replenishes the losses and adds only little (if any) to population increase.

In most lakes only a few resources are limiting for phytoplankton growth: light, one to three of the "classic" nutrients (P, N, Si) and maybe one or the other trace element. In most cases the number of limiting resources is clearly less than ten. This contrasts to several tens to hundreds of species coexisting in even small parcels (ml) of water in a spatially quite unstructured environment. This discrepancy was termed "paradox of the plankton" by Hutchinson (1961). The attempts to "solve Hutchinson's paradox" have not only given rise to Tilman's competition theory, they are also important in relation to the general ecological question of the relationship between diversity and the constancy or variability of the environment. The classic view is that diversity increases with environmental constancy and with successional "maturity". This view drew its support from an idealized image of the constancy of tropical climate and from a premature identification of some tropical communities (rain forests, corral reefs) with the preconceived concept of the "climax stage". Apparently the classic view contradicts the competitive exclusion principle and is not universally supported by empirical data from late successional forests. Still it persists in a number of textbooks on ecology (e.g., Margalef 1977). In contrast, Connell's (1978) "intermediate disturbance hypothesis" postulates that unperturbed communities would develop towards low diversity and that disturbances of intermediate frequency and intermediate intensity are needed to maintain diversity.

Competitive exclusion needs time. In a chemostat competition experiment with natural phytoplankton from Lake Constance at a Si:P (the two limiting nutrients) ratio of 20 and a daily renewal rate of 30%, the Si-limited diatom *Synedra* and the P-limited green alga *Koliella* excluded all other species (Fig. 3, left). But it took about 2 weeks until the last "loser" started to decline and it took about 3 to 4 weeks until the winning couple made up 95% of the total biomass. This time demand is quite uniform across many experiments with algae from different lakes and under quite different conditions: 14.4 ± 3.1 days (SD) until the last loser declines and 23.3 ± 5 days (SD) until the winners make up 95% (Sommer 1989). Thus, for quite a long time the phytoplankton community carries a "memory" of species which are now losing competitors but may have had better conditions earlier. As long as they are not completely excluded a reversal in competitive conditions may give them another chance. This is shown in the right

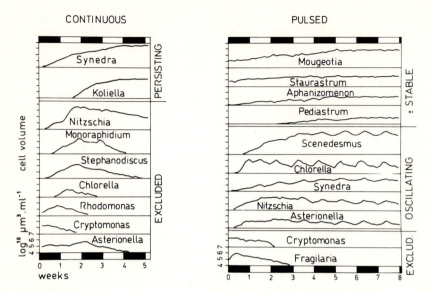

Fig. 3. Continuous flow competition experiments with natural phytoplankton; Si:P ratio 20:1 (molar), dilution rate: 0.3 days^{-1}; biomass (cell volume) of most important species in log scale. *Left* Chemostat experiments (Si and P added continuously as a component of the flow-through medium); *right* Pulsed experiments (Si and P added once per week in a pulse) (After Sommer 1985)

half of Fig. 3. If P and Si are administered in weekly pulses instead of being a component of the continuous flow medium, some species decline when Si or P become scarce during the second half of the week but recover when the nutrient pulse reverts the competitive arena to their favor (*Chlorella* and *Scenedesmus* because of the P-pulse, *Synedra*, *Asterionella*, and *Nitzschia* because of the Si-pulse). Some other species (*Mougeotia*, *Staurastrum*, *Aphanizomenon*, *Pediastrum*) do not seem to respond very much to the nutrient pulse, either because they build up an internal storage pool or because they can also obtain enough nutrients during the impoverished period. In total, nine species persisted in the pulsed regime as opposed to two species in the constant regime. Connell would have called the weekly nutrient pulses "intermediate disturbances". But which frequency of disturbances would be best for the maintenance of diversity?

This was tested by another series of experiments (Gaedeke and Sommer 1986). Instead of a continuous flow-through, cultures were diluted discontinuously at regular intervals from 1 to 14 days. The magnitude of the dilution was adjusted to give the same long-term dilution rate to all cultures (0.3 days^{-1}) and thus an average generation time of 2.3 days for algae in equilibrium with the losses by dilution. At dilution intervals below one mean generation time diversity did not increase above the level which can be attained theoretically with two limiting resources at steady state. Only dilution intervals in excess of one generation time were fertile in terms of diversity (Fig. 4). A peak of diversity was attained at a dilution interval of 7 days, or ca. three generation times.

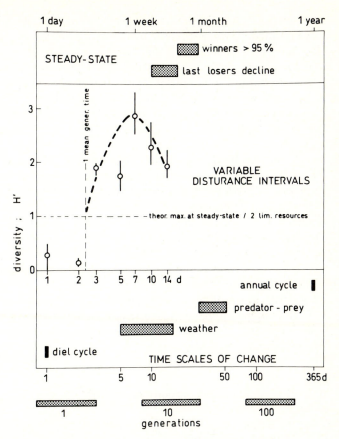

Fig. 4. Phytoplankton diversity and the "intermediate disturbance" hypothesis; time is given in a logarithmic scale. *Top panel*: time needed for competitive inclusion in steady-state experiments. *Middle panel*: diversity (Shannon-Weaver H') of phytoplankton in competition experiments with variable intervals between dilutions. *Bottom panel*: time scales of disturbances in situ

How does this relate to the nature and frequency of disturbances encountered by phytoplankton in situ? The mean generation time in those experiments is quite realistic for summer conditions and can be attained by resource-limited, small and medium algae, and resource-saturated large ones. The dilution events have a dual impact: they decrease the population density and they replenish exhausted resources. As a consequence, per capita resource availability is increased. The same happens if the depth of the mixed layer is increased by destratification: epilimnic phytoplankton populations are diluted and fresh nutrients are imported from deeper waters. This is also the case if grazing rates increase, algae decline, and nutrient excretion by zooplankton increases.

Both types of "disturbances" occur at various time scales. The change between stratification and destratification has an annual, an intermediate, and a diel cycle. The annual cycle comprises one (monomictic) or two (dimictic lakes)

periods of complete circulation of the water and stratification in the time between. The intermediate cycle is caused by the change of weather conditions during the stratified period: cooling and wind events increase the mixing depth, warming and absence of wind enforce stratification. Of course, the interval between such partial destratification events is highly irregular but in the temperate zone there is a dominant time scale of about 5 to 15 days (Harris 1986). The diel cycle is caused by daytime warming of the surface and buildup of a week stratification within the epilimnion and nighttime cooling and convective mixing of the epilimnion.

Zooplankton grazing also has an annual, an intermediate, and a diel scale. The diel change is behavioral (vertical migration with a grazing peak at night; Lampert and Taylor 1985). The intermediate and the annual scale of variation are demographic. Superimposed to the annual cycle of abundance zooplankton, together with phytoplankton, often undergo predator-prey oscillations. For *Daphnia* (the most important grazer in many lakes) 30 to 50 days is the characteristic wavelength of such oscillations (McCauley and Murdoch 1987).

The comparison of the experimental results with the time scale of natural disturbances (Fig. 4) indicates that partial destratification within the stratified period (the intermediate scale of physical disturbance) is most effective in terms of promoting phytoplankton diversity and that the diel variation of conditions is quite ineffective. However, this conclusion is quite preliminary, given the paucity of experimental results.

4 Is Phytoplankton Succession Convergent?

Phytoplankton competition experiments have been highly replicable worldwide. For instance, whenever natural inocula have been used and whenever Si:P ratios have been sufficiently high to exclude Si-limitation, members of the genus *Synedra* became dominant (summarized in Sommer 1989), irrespective of its initial abundance in the inoculum and of the geographic origin of the inoculum. In the long run, physiologically superior competitors overcome any disadvantage imparted by initial rareness and physiologically inferior competitors lose any advantage imparted by initial dominance. This has been shown by mutual invasion experiments where a small inoculum of an invader species was added to an established equilibrium population of another species in chemostats (Tilman and Sterner 1984). Whenever the invader was the physiologically superior competitor it started to increase immediately after invasion and the population of the resident species started to decline within the first 2 days.

Thus, it is obvious that phytoplankton succession is convergent in competition experiments. Only the physical and chemical conditions and the physiological abilities and requirements of the competing species, but not the initial abundance of the competitors, determine the equilibrium state. As stated above, a 95% approximation to this equilibrium state is usually reached after approximately 23 days. However, such experiments do not take into account the interactions with higher trophic levels. Grazing zooplankton may seriously influence the succession of phytoplankton by selective removal of more vulnerable algae (Porter 1973; Gliwicz 1975) and by selective excretion of limiting nutrients (Sommer 1988; Sterner 1989). In the presence of further trophic levels convergence of phytoplankton succession may be unaffected, it may be delayed, or it may be broken, because such a more complex system might have alternative stable states (or stable cycles).

A preliminary answer for three-level systems (abiotic resources — phytoplankton — zooplankton) can be derived from microcosm experiments where different initial mixtures of phytoplankton and zooplankton are cultivated under identical physical and chemical conditions (Figs. 5 and 6; Sommer, unpubl.). The different inoculum compositions were obtained by mixing phytoplankton and zooplankton samples from three different lakes in different proportions (e.g. 10:1:1; 1:10:1; 1:1:10; 1:1:1). Three experiments were conducted under sedimentation-intensive (weak stirring) conditions and two under avoidance of sedimentation. Within experiments, convergence of phytoplankton species composition (as expressed by the similarity index of Bray and Curtis 1957) was attained after 7 to 9 weeks. In the sedimentation-intensive experiments the final phytoplankton assemblage consisted of the filamentous green alga *Mougeotia*

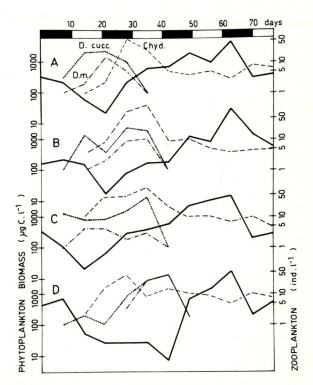

Fig. 5. Phytoplankton biomass (*thick line*) and population density of dominant zooplankton species (*Daphnia cucculata*, ...; *D. magna*, -.-.-.; *Chydorus sphaericus*, ----) in a microcosm experiment with different inoculum compositions under sedimentation-intensive conditions (Sommer, unpubl.)

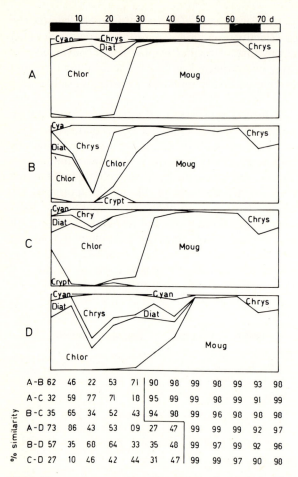

Fig. 6. Taxonomic composition (Cyanophyta, Cryptophyta, Chrysophyceae, diatoms, Chlorophyta excl. *Mougeotia, Mougeotia*) in the experiment shown in Fig. 5. *Numbers* at the bottom give the pairwise similarity indices between the different microcosms. The inocula were different mixtures from the lakes Schöhsee, Kellersee and Ciro Ber Binnensee (**A** 1:1:1; **B** 10:1:1; **C** 1:10:1; **D** 1:1:10)

thylespora (poorly edible for zooplankton, good competitor for P; see Sommer 1983, 1988, 1989) accompanied by a small biomass of edible nanoplankton. In the experiments without sedimentation the final assemblage consisted of diatoms of the family Fragilariaceae (*Synedra acus, Asterionella formosa, Fragilaria crotonensis,* very good competitors for P, poorly edible for zooplankton) accompanied by a small biomass of nanoplankton.

The final phytoplankton assemblage showed symptoms of strong P-limitation (C:P ratio is the biomass > 500:1). The finally dominant zooplankton species (the Cladocera *Bosmina* spp., *Chydorus sphaericus,* the rotifer *Lecane* sp.) probably did not feed on the dominant large algae but only on the "undergrowth flora" of small algae and bacteria (Geller and Müller 1981). Contrary to *Daphnia*, which was replaced quite early during the experiments, their feeding is also not impeded by large algae (Gliwicz 1977). Interestingly, complete taxonomic convergence of zooplankton was not required before the phytoplankton species composition converged. In some experiments zooplankton converged taxonomically before the termination, in some it did not. Obviously, the zooplankton species making up the final assemblage are functionally equivalent from the view point of phytoplankton. *Chydorus* and *Bosmina* were equivalent at a 1:1 basis, whereas about 20 to 50 individuals of *Lecane* were equivalent to one Cladoceran individual.

In conclusion, within the experimental systems presented here, phytoplankton succession was convergent, although the presence of a further trophic level increased the time needed for convergence about twofold. Analogous experiments with one further trophic level (zooplankton-feeding fish) cannot be performed at the scale of microcosm laboratory experiments. At present, it is not yet clear to what extent the influence of fish on zooplankton is further propagated to phytoplankton. If there is a strong top-down control as suggested by the trophic cascade theory (Carpenter et al. 1985) neither a further retardation of the convergence of phytoplankton succession nor the existence of multiple stable states can be excluded. If top-down effects of fish are dampened when cascading down the trophic chain (McQueen and Post 1988), no large effect of fish is to be expected.

5 Do Phytoplankton Conform to the Mosaic-Cycle Theory?

It has been shown that phytoplankton succession is quite regular. Preliminary evidence also suggests that phytoplankton succession is convergent under similar external conditions. Late successional stages in the absence of external disturbances are characterized by low diversity. Disturbances at intermediate frequencies (interval of a few generation times) seem to be most effective in maintaining phytoplankton diversity. In conclusion, phytoplankton seems to conform to Connell's (1978) "intermediate disturbance hypothesis". The relationship to the "mosaic-cycle theory" (Remmert 1985, this Vol.) must still be explored.

Compared to the intermediate disturbance hypothesis, the mosaic-cycle theory is the more restrictive concept. The mosaic-cycle theory agrees with the intermediate disturbance hypothesis about the importance of disturbances in the maintenance of diversity. In addition, the mosaic-cycle theory postulates that disturbances are a necessary consequence of the inability of late successional ("climax") stages to reproduce themselves locally, because adult dominant organisms prevent the regrowth of conspecific juveniles. Only local dieback of senescent cohorts permits the renewed start of a local succession towards the same final stage. The intermediate disturbance hypothesis makes no prediction as to whether late successional stages are self-sustainable or not.

Phytoplankton competition experiments do not support the inability of competitively dominant populations to reproduce themselves locally. Artificial communities produced by competition in continuous cultures seem self-sustainable and show no sign of "senescence". It may, however, be questioned whether the continuity of conditions, as realized in a chemostat, would be valid for natural conditions, even in the absence of external disturbances. It can be easily conceived that phytoplankton communities alter the resource basis in a way that the currently dominant species would lose their competitive advantage. Diatom plankton is such a case during the stratified period. Contrary to N and P, silicate is not recycled by grazing zooplankton (Sommer 1988). Most of the particulate diatom Si will be lost by sedimentation. Thus, consumption of silicate by a growing diatom population will reduce the Si:P ratio and thus change the competitive arena to the disadvantage of the diatoms. In conclusion, summer diatom stages are not self-sustainable.

On the other hand, there is currently no evidence that typical late summer stages of phytoplankton succesion (dinoflagellates, Cyanobacteria) lack the ability of reproducing themselves under unchanged external conditions. Usually, the frequency of external disturbances excludes a clear judgement, but occasionally very stable summers in wind-sheltered lakes illustrate how long dominance of late successional species can persist. The summer of 1989 in Plußsee (FRG) was such a case (Fig. 7). In May a steep temperature gradient built up at 4 to 6 m depth. This thermocline strongly restricted vertical mixing of water. It was sufficiently stable to withstand strong cooling of the surface water in early July, mid-July, early August and early September. During that period of physical stability, the large dinoflagellate *Ceratium hirundinella* and a smaller amount of its congener *C. furcoides* continuously increased and nearly completely replaced all other species. Consequently, the diversity index dropped to extremely low levels. During that period, *Ceratium hirundinella* was moderately N-limited and reproduced with a doubling time of 2.5 to 4 days. Because of unexplained losses, the actual population increase from mid-May till the peak in mid-September was about one doubling per week. The period of continued decrease lasted about 40 generations, the period of dominance in terms of biomass lasted about 30 generations. This is much longer than the longevity of any so-called climax stage in forest succession. In accordance with the principle of competitive exclusion, the diversity of phytoplankton declined to extremely low values as compared to spring and to the starting conditions in early summer. The diversity of the phytoplankton community could not recover before the onset of autumn destratification when the dominant *Ceratium* population was almost completely eliminated by conversion into cysts which were lost from the mixed layer by sinking.

In conclusion, the seasonal succession of phytoplankton seems to be more in line with the traditional succession theory than with the mosaic-cycle theory. Early and mid-successional stages are intrinsically transient, while the final stages seem to be self-sustainable. It is only the external cycle of climatic and hydrographic conditions which forces plankton succession back to the starting point each winter.

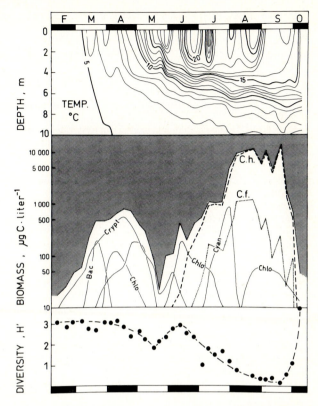

Fig. 7. Phytoplankton succession in Plußsee from February to October 1989. *Top panel*: temperature stratification. *Middle panel*: biomass of total phytoplankton and of taxa (log scale): *Crypt* Cryptophyceae; *Bac* Bacillariophyceae; *Chlo* Chlorophyta; *Cyan* Cyanobacteria; *dashed line*: *Ceratium hirundinella*; *dotted line*: *C. furcoides*. *Bottom panel*: Diversity index (Shannon-Weaver)

References

Bray JR, Curtis TJ (1957) An ordination of the upland forest communities of southern Wisconsin. Ecol Monogr 27:325–349

Carpenter SR, Kitchell JF, Hodgson JR (1985) Cascading trophic interactions and lake productivity. BioScience 35:634–639

Connell JH (1978) Diversity in tropical rain forests and coral reefs. Science 199: 1304–1310

Gaedeke A, Sommer U (1986) The influence of the frequency of periodic disturbances on the maintenance of phytoplankton diversity. Oecologia 71:25–28

Geller W, Mueller H (1981) The filtration apparatus of Cladocera: filter mesh-sizes and their implications on food selectivities. Oecologia 49:316–321

Gliwicz ZM (1975) Effect of zooplankton grazing on photosynthetic activity and composition of phytoplankton. Verh Int Verein Limnol 19:1480–1497

Gliwicz ZM (1977) Food-size selection and seasonal succession of filter-feeding zooplankton in a eutrophic lake. Ekol Polsk 25:179–225

Harris GP (1986) Phytoplankton ecology. Chapman and Hall, London

Hutchinson GE (1961) The paradox of the plankton. Am Nat 95:137–147

Lampert W, Taylor BE (1985) Zooplankton grazing in a eutrophic lake: implications of diel vertical migration. Ecology 66:68–82

Margalef R (1977) Ecologia. Omega, Barcelona

McCauley E, Murdoch WW (1987) Cyclic and stable populations: plankton as paradigm. Am Nat 129:97–121

McQueen DJ, Post JR (1988) Cascading trophic interactions: uncoupling at the zooplankton-phytoplankton link. Hydrobiologia 159:227–296

Porter KG (1973) Selective grazing and differential digestion of algae by zooplankton. Nature (Lond) 244:179–180

Porter KG (1977) The plant-animal interface in freshwater ecosystems. Am Sci 65:159–170

Remmert H (1985) Was geschieht im Klimax-Stadium? Naturwiss 72:505–512

Reynolds CS (1980) Phytoplankton assemblages and their periodicity in stratifying lake systems. Holarct Ecol 3:141–159

Reynolds CS (1984) Phytoplankton periodicity: the interaction of form, function and environmental variability. Freshwater Biol 14:111–142

Reynolds CS (1988) Functional morphology and the adaptive strategies of freshwater phytoplankton. In: Sandgren CD (ed) Growth and survival strategies of freshwater phytoplankton. Cambridge Univ Press, Cambridge, pp 388–433

Reynolds CS (1989) Physical determinants of phytoplankton succession. In: Sommer U (ed) Plankton ecology: succession in plankton communities. Springer, Berlin Heidelberg New York Tokyo, pp 9–56

Reynolds CS, Wiseman SW (1982) Sinking losses of phytoplankton in closed limnetic systems. J Plankton Res 4:489–522

Reynolds CS, Thompson JM, Ferguson AJD, Wiseman SW (1982) Loss processes in the population dynamics of phytoplankton maintained in closed systems. J Plankton Res 4:561–600

Reynolds CS, Wiseman SW, Godfrey BM, Butterwick C (1983) Some effects of artificial mixing on the dynamics of phytoplankton in large limnetic enclosures. J Plankton Res 5:203–234

Sommer U (1983) Nutrient competition between phytoplankton species in multispecies chemostat experiments. Arch Hydrobiol 96:399–416

Sommer U (1984) Sedimentation of principal phytoplankton species in Lake Constance. J Plankton Res 6:1–15

Sommer U (1985) Comparison between steady state and non-steady state competition: experiments with natural phytoplankton. Limnol Oceanogr 30:335–346

Sommer U (1986) The periodicity of phytoplankton in Lake Constance (Bodensee) in comparison to other deep lakes. Hydrobiologia 138:1–7

Sommer U (1987) Factors controlling the seasonal variation in phytoplankton species composition. A case study for a nutrient rich, deep lake. Progr Phycol Res 5:123–178

Sommer U (1988) Phytoplankton succession in microcosm experiments under simultaneous grazing pressure and resource limitation. Limnol Oceanogr 33:1037–1054

Sommer U (1989) The role of competition for resources in phytoplankton succession. In: Sommer U (ed) Plankton ecology: succession in plankton communities. Springer, Berlin Heidelberg New York Tokyo, pp 57–106

Sommer U, Gliwicz ZM, Lampert W, Duncan A (1986) The PEG-model of seasonal succession of planktonic events in fresh waters. Arch Hydrobiol 106:433–471

Sterner RW (1989) The role of grazers in phytoplankton succession. In: Sommer U (ed) Plankton ecology: succession in plankton communities. Springer, Berlin Heidelberg New York Tokyo, pp 107–170

Tilman D (1982) Resource competition and community structure. Princeton Univ Press, Princeton

Tilman D, Sterner RW (1984) Invasions of equilibria: test of resource competition using two species of algae. Oecologia 61:197–200

Mosaic-Like Events in Arid and Semi-Arid Namibia

H.H. Berry[1] and W.R. Siegfried[2]

1 Introduction

The mosaic-cycle hypothesis, originally proposed by Aubreville (1938), has been recently reviewed by Remmert (1987). In essence, Aubreville's hypothesis predicts that if a tree or group of trees dies in a natural, tropical forest, the empty space will be occupied by new pioneer plants which will be replaced by a succession of vegetation phases ending in the original "climax" vegetation. In this manner cycles of varying "mosaic" phases are created. Remmert (1987) extended this hypothesis to all natural systems and posed, inter alia, the following questions:

1. "What agents drive the cycles? Is it the longevity of the key organisms or have we to look for fungi, microbes and animals?"
2. "What is the size of the mosaic stones and what determines the size?"

Apart from the study by Weiner and Gorecki (1982) on rodents in the arid steppe of central eastern Mongolia, apparently the mosaic-cycle hypothesis has not been considered in the context of desert and semi-desert environments. We address Remmert's (1987) questions via two case histories, involving relatively long-term data sets and anthropogenic influences in arid and semi-arid Namibia. Our case histories are based on the behaviour of large mammals and the ostrich *Struthio camelus* and the influence of modern man over periods of up to 30 years in the arid Namib-Naukluft Park and in the semi-arid Etosha National Park. We briefly describe these two areas and then organize the narrative under primary driving events for, and primary effects of, the selected biota, interpreting these features in the context of the mosaic-cycle hypothesis.

[1] Namib-Naukluft Park, Directorate of Nature Conservation, P.O. Box 1592, Swakopmund 9000, Namibia, Africa
[2] Percy FitzPatrick Institute, University of Cape Town, Rondebosch 7700, South Africa, Africa

H. Remmert (Ed.)
Ecological Studies Vol. 85
© Springer-Verlag Berlin Heidelberg 1991

2 Study Areas

2.1 Namibia

Our observations were made in Namibia (17°–29°S, 12°–21°E) in southwestern Africa. Namibia covers some 824 000 km² of which 28% can be classified as arid (< 150 mm/annum) and 69% as semi-arid (150–600 mm/annum; Tinley 1975). This paucity of rainfall (Fig. 1) is a consequence of two major factors: a South Atlantic high-pressure atmospheric cell is the dominant, off-shore, climatic feature, producing "subsidence inversion" which limits convection (Preston-Whyte et al. 1977); and, in the Benguela Current region, cold, upwelled and nutrient-rich, coastal waters enhance the stability of the lower air layer and limit its moisture capacity. Furthermore, precipitation by rainfall is sporadic and unpredictable, resulting in accompanying fluctuations in plant and animal abundance.

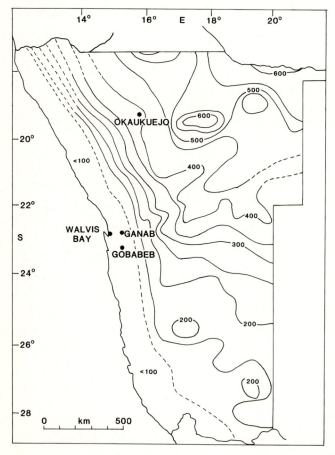

Fig. 1. Isohyetal map of Namibia (source: Weather Bureau, Windhoek, 1980)

2.2 Namib-Naukluft

The Namib-Naukluft Park is located in the central Namib desert, and includes the Namib Research Institute which has stations at Gobabeb (23°34 S, 15°03 E) and Ganab (23°09 S, 15°33 E; Fig. 2). The coastal, gypsum-gravel plains of the Namib desert are usually devoid of vegetation, except for a few hardy perennial plants in the dry watercourses and a diverse lichen flora accompanied by scattered dwarf shrubs on the gypsum surfaces (Seely 1987). This zone falls within the fog belt, which extends inland from the Atlantic Ocean to a distance of more than 100 km (Lancaster et al. 1984). Rainfall increases from 15 mm/annum on the coast to 65 mm/annum 110 km inland, and the vegetation is characterized by intermittent annual grassland with perennial plants largely restricted to watercourses (Seely 1978).

Two major moisture sources dominate this region: fog and rain. Whereas fog is a relatively regular, predictable phenomenon, which precipitates sufficient moisture (31 mm/annum at Gobabeb) to ensure survival of the greatest diversity of invertebrates in any desert in the world, rainfall is rare and unpredictable

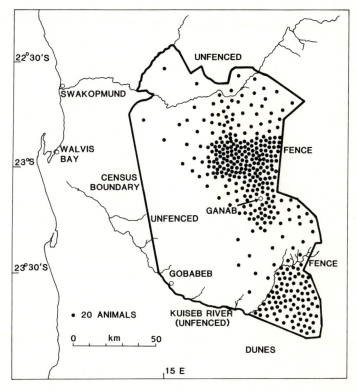

Fig. 2. Map of the Namib-Naukluft Park, giving combined densities of ostrich, springbok, gemsbok and Hartmann's zebra during an aerial census in October 1982. The area of greatest animal density in the vicinity of Ganab corresponds to the open gravel plains

(Pietruszka and Seely 1985; Seely 1987). Significant rainfall events may be viewed as erratic pulses of water which result in brief flushes of highly nutritious ephemeral grasses in a desert whose barren nature normally limits the populations of large herbivores (Seely and Louw 1980).

2.3 Etosha

Etosha National Park lies within the 300-mm isohyet in the west and the 500-mm isohyet in the east (Fig. 3). Rainfall is the only measurable form of precipitation at Etosha. Three climatic periods have been identified during the year: wet and hot (January-April), dry and cool (May-August) and dry and hot (September-December) (Berry 1980). One of the park's major physical features is a vast (4690 km²) salt-pan that has been described as "saline desert with dwarf shrub savanna fringe" by Giess (1971) in his subdivision of the major vegetation zones of the country. It is on the fringe of this salt-pan that a sweet, short grassveld on calcareous soil attracts abundant populations of large ungulate grazers following good rains (Berry and Louw 1982). Fire does not play a significant role in the maintenance of these short grasslands, although it is a regular feature in the surrounding savanna vegetation (Siegfried 1981).

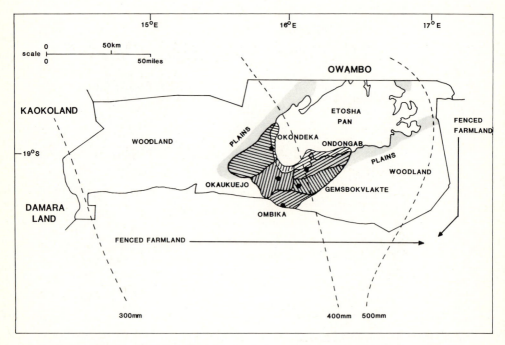

Fig. 3. Map of the Etosha National Park. The *hatched* zones indicate major areas for lions, wildebeest, zebra and anthrax

3 Primary Events and Effects

3.1 Namib-Naukluft

The rainfall records at Gobabeb, measured since 1962, show a 13-year period (1962–75) in which 15.4 mm/annum was the average (Fig. 4), which is lower than the long-term mean of 19.8 mm. An exceptional rainfall of 106.7 mm was measured in 1976 followed by 84.4 mm 2 years later. The subsequent 11 years (1978–89) were dry, with an average of only 12.0 mm/annum. At Ganab, rainfall since 1967 has followed a sequence similar to Gobabeb: the annual precipitation averaged at least 183 mm (measurements incomplete) in the 3-year period 1974–76), with a peak of at least 160 mm in 1974 and at least 355 mm in 1976 (Fig. 5). This is considerably higher than the long-term mean of 70.4 mm. Similarly to Gobabeb, Ganab's rainfall decreased sharply thereafter, averaging only 47.2 mm/annum for the 13-year period 1977–89.

The effect of unusually high rainfall events on the abundance of the area's four principal large herbivores in the gravel-plains' grasslands is obvious (Figs. 4 and 5). These animals are the ostrich, springbok *Antidorcas marsupialis*, gemsbok *Oryx gazella* and Hartmann's zebra *Equus zebra*. At both Gobabeb and Ganab, peaks in large herbivore populations were reached 3 years after the high rainfall of 1976. Subsequently, as rainfall decreased, herbivore numbers fell sharply over a period of 10 years. This reduction was promoted by mortality, when an average of nine fresh carcasses per day were found (Nel 1980), recruitment failure, and

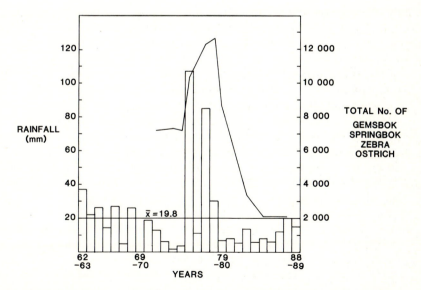

Fig. 4. Rainfall records (histogram) for Gobabeb (1962/63–1988/89), in the Namib-Naukluft Park, calculated from 1 July to 30 June each year, in relation to total numbers of four herbivore species (ostrich, springbok, gemsbok, Hartmann's zebra) for the period 1972–88

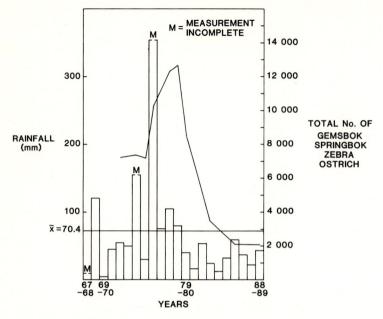

Fig. 5. Rainfall records (histogram) for Ganab (1967/68–1988/89), in the Namib-Naukluft Park, calculated from 1 July to 30 June each year, in relation to total numbers of four herbivore species (ostrich, springbok, gemsbok, Hartmann's zebra) for the period 1972–88

movement away from the area as these large, mobile animals retreated to areas of more stable productivity (Boyer 1988). The one large predator of any consequence in this area is the spotted hyaena *Crocuta crocuta*. However, only a small fraction of its diet consisted of prey taken from the plains; the overwhelming majority being from the adjoining Kuiseb riverbed and the sand-dune fields in the south (Tilson et al. 1980). Consequently, neither predation nor disease had a significant effect on the abundance of large herbivores during this period.

Of the four herbivorous species considered here, the ostrich increased more rapidly after the increase in rainfall; springbok, gemsbok and Hartmann's zebra, requiring 2 and 3 years, respectively, to reach peak numbers (Fig. 6). All four species decreased with equal rapidity. Linear regressions of these population curves were drawn against rainfall measured at Ganab, where the majority of animals congregated. The regression for the ostrich confirmed the species' predictable response to rainfall as shown in Fig. 7 ($y = 9.21x + 576.92$) with a good fit to the data $r = 0.91$. The regressions for springbok, gemsbok and zebra, however, were much weaker due to their lagged responses and the coefficients of determination in all cases ($r = <0.1$) gave a poor fit to the data. When the numbers of the four species were totalled, the regression showed no predictable linear relationship against rainfall ($y = 81.22x + 5650.33$) and the fit to the data was unacceptable ($r = 0.23$) as shown in Fig. 7.

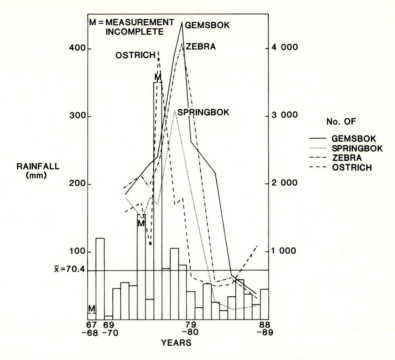

Fig. 6. Numbers of four large herbivore species (ostrich, springbok, gemsbok and Hartmann's zebra) and rainfall (histogram) in the Namib-Naukluft Park (1967/68–1988/89)

If rainfall was the single natural primary event or agent for the biotic changes reported here, then the influence of fences and artificial waterholes should also be examined. Four borehole-fed waterholes were established on the desert plains after the eastern perimeter of the Namib-Naukluft Park was fenced in 1967. When food demands by herbivores exceed the desert's ephemeral ability for primary production, mobility to alternative feeding sites or drought-induced mortality are the only alternatives (Hamilton et al. 1977). Fencing and artificial water supplies probably increased the mortality of herbivores by retarding their emigration. Weak individuals began dying in the vicinity of waterholes and those animals which attempted to emigrate were hindered by the fence. However, the fence is not completely effective and cases do occur in which animals find or force their way through it (J. Lenssen, pers. commun. of the Chief Nature Conservation Officer, Namib-Naukluft Park 1989).

In summary, the immediate consequence of the primary agent driving the observed "cycle", namely rainfall, was a widespread, but brief, "explosion" of nutritious ephemeral grass on the gravel plains, followed by herbivore immigration and production. The standing crop of the grass is positively correlated with rainfall (Seely 1978). Subsequently, food stocks were depleted, causing high mortality rates among the animals, whilst an unknown portion of the surviving

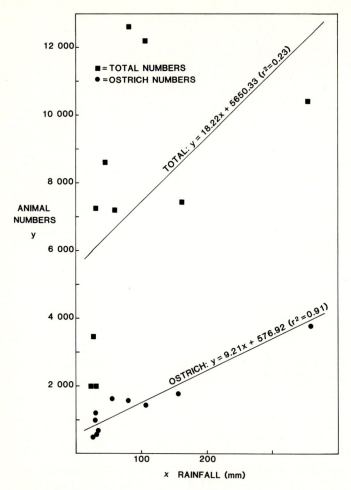

Fig. 7. Relationships between the combined total populations of four herbivore species (ostrich, springbok, gemsbok and Hartmann's zebra) and rainfall, and ostrich and rainfall in the Namib-Naukluft Park (1967/68–1988/89)

populations emigrated across the park's eastern fence or southwards into the sand-dune fields.

3.2 Etosha

The rainfall record, kept at Okaukuejo in the Etosha National Park since 1956, fluctuated greatly around the annual mean of 354 mm (Fig. 8). From 1956 to 1965 precipitation averaged 320 mm/annum which is 10% below the 33-year mean. Subsequently, the period 1966–79 was a "wet" phase, with annual precipitation

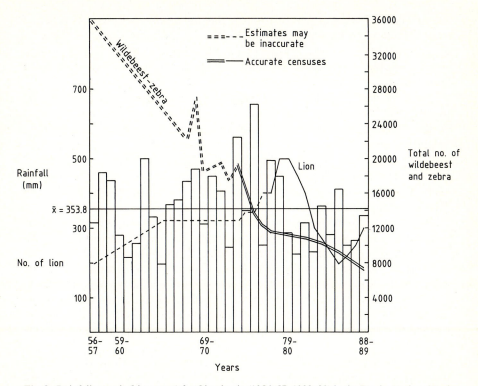

Fig. 8. Rainfall records (histogram) for Okaukuejo (1956/57–1988/89) in the Etosha National Park, calculated from 1 July to 30 June each year, in relation to numbers of lions (on the rainfall axis) and wildebeest-zebra (on the right-hand axis)

averaging 417 mm (18% above mean) and reaching a maximum of 678 mm (92% above mean) during 1976. The period 1980–89 was comparatively dry, with annual precipitation averaging 297 mm (16% below mean) and decreasing to 223 mm (37% below mean) during 1981. In spite of the clear wet and dry periods, we were unable to find significant statistical correlations between amounts of rainfall and any of Etosha's large mammals.

Prior to 1970 the large mammalian herbivores and probably their attendant predators migrated away from Etosha during the 8-month dry season each year. They returned briefly for the 4 months (January–April) in which rainfall is sufficient to produce the highly nutritious annual grasses which constitute their preferred grazing (Berry 1980). In 1970 the 850-km perimeter of Etosha National Park was fenced to a height of 2.5 m, effectively curbing the movements of most large animals, with the notable exception of the elephant *Loxodonta africana*. The animals remaining within the park were further subjected to two other artificial factors in the form of 54 waterholes, and several hundred borrow pits for gravel used in road-building. The artificial waterholes facilitated the expansion and stabilization of the lion *Panthera leo* population, whilst the latter apparently created favourable conditions for anthrax *Bacillus anthracis*, the most viable and virulent bacterium known (Turnbull 1989), and a disease fatal to herbivores (Ebedes 1977; Berry 1981). (Most of the waterholes have subsequently been closed by the park's managers; the number of operative artificial waterholes being 19 in 1989.)

Anthrax was probably accidentally introduced into Etosha via infected domestic livestock during 1955–60 and rapidly became epidemic (Ebedes 1977). Ebedes suggested that this had occurred mainly through optimum conditions created by the alkaline nature of the gravel pits used in road-building. Subsequently, it has been proposed (Turnbull et al. 1989) that waterholes and gravel pits are not sites of propagation of anthrax. Nevertheless, whatever the source of infection, anthrax has been established as a major cause of mortality in herbivores at Etosha (Turnbull 1989).

Anthrax outbreaks appear to be predictably linked to the wet season in the case of wildebeest *Connochaetes taurinus* and zebra *Equus burchellii* (Ebedes 1977), whereas elephant mortalities resulting from anthrax occur in the dry season. This apparent paradox may be explained by the observation (Berry, unpubl. data) that elephants browse *Acacia* species during dry periods. Thorns may cause lesions in their mouths, making them particularly susceptible to the passage of anthrax when drinking from infected waterpoints. This susceptibility may also be enhanced by the nutritional stress which elephants are prone to under dry conditions.

The combined effects of fencing, artificial waterholes and abnormal levels of anthrax resulted in dramatic changes in several species. For example, wildebeest decreased from 25 000 in 1954 to 2500 individuals in 1978, staying at approximately that level until the present. Similarly, zebra decreased from 22 000 in 1969 to 5000 individuals in 1988. Whilst these two herbivores appear to have been adversely affected by reductions in rainfall and increases in man's influence, the lion population has responded differently (Fig. 8). The lion population increased steadily from 200 to 500 individuals between 1954 and 1979. The increase in lion abundance led to an abnormally high predator-prey ratio (1:150) in Etosha (Berry 1981). Thereafter, the population decreased to 200 individuals by 1986. Since then it has again increased: the most recent (1989) estimates include a minimum of 266 and a maximum of 341 lions (Stander 1990).

One reason for the lion's resilience is its ability to behave as a predator or a scavenger with equal ease. Furthermore, lions have developed an immunity to anthrax (Turnbull et al. 1989), which herbivores have not, and so are able to feed on anthrax carcasses without any adverse effects. In Etosha, this phenomenon resulted in a decrease in the incidence of the lion's hunting, while the reduced migration of herbivorous prey enhanced the carnivores' recruitment. The presence of permanently available artificial waterholes throughout the park facilitated the lions' colonization of large new areas.

The advent (ca. 1980) of the worst drought in recorded history in Etosha caused the already reduced wildebeest and zebra populations to shift from the then unproductive grassy plains into the adjoining bushland. This in turn forced prides of lions to shift their territories into finite areas, where living space was limited. Since lion-pride territories disperse animals (Schaller 1972), the displaced lions of Etosha were wedged between a drought-stricken plains system and adjoining cattle farms. Inevitably, lions began trespassing onto farmland and killing large numbers of domestic livestock (one male was responsible for killing more than 100 head of cattle, horses, donkeys and goats in 2 years before being shot). At least 317 lions were destroyed on farms in 10 years, representing a significant proportion of Etosha's population (Table 1). During the same period, a total of 20 lion carcasses was found inside Etosha but, because freshly dead lions are seldom located, the cause (rabies) of death could only be determined in three of these.

Table 1. Minimum number of lions reported destroyed while trespassing on farms adjoining Etosha National Park (1978–87) according to Berry (1987)

Year	Population in park	No. trespassers destroyed	Percentage population destroyed
1978	400	37	9
1979	500	11	2
1980	500	25	5
1981	450	42	9
1982	400	84	21
1983	300	39	13
1984	270	31	11
1985	230	25	11
1986	200	12	6
1987	220	11	5

In summary, as in the Namib-Naukluft case, the primary agent driving changes, and perhaps cycles, in Etosha's populations of large herbivores is sporadic rainfall. Superimposed on this there is a significant anthropogenic influence which affects interactions between the large herbivores and their predators. As large herbivores decreased in the system, lions and anthrax bacteria flourished quasi-symbiotically. The bacteria promoted a greatly increased supply of food for the lions which, in turn, aided the dispersal of anthrax by carrying its sporulated form in their mouths, claw sheaths and other body parts to infect new sources. The favourable phase for lions was, however, short-lived, and as their food supply diminished in response to prolonged drought and reduced productivity of the grassland, they were forced onto farmland with fatal consequences.

4 Discussion

Both of the case histories we have presented have certain predictive capabilities for the mosaic-cycle hypothesis, but most of the predictions cannot (yet) be demonstrated statistically. In the arid Namib-Naukluft area "wet" phases may be

of relatively brief duration (possibly 2–3 years) and dry phases may last at least
11–13 years, based on present knowledge. To speak of quasi-regular, alternating
cycles of "wet" and "dry" periods in this area would be premature. It is, however,
clear that it is the presence or absence of significant rainfall that primarily drives
the system and shapes its phases. These phases, in terms of size, duration,
structure and functioning, could be analogous to "mosaic stones" (sensu Remmert 1987).

In Etosha rainfall is probably also the primary agent driving the system, with
successive wet and dry phases of 10–14 years on record. Although Tyson (1986)
identifies an oscillation of 9 years between dry and wet periods in southern Africa,
it is not possible at present to say whether the rainfall in Etosha fits this pattern or
not. Despite years of good rainfall and grass production, sustained, spectacular
decreases of wildebeest and zebra are still continuing. These decreases are being
promoted by anthrax, predators and fences. In contrast, anthrax and lions are still
flourishing in Etosha, the former by expanding its hosts to include elephant and
black rhinoceros *Diceros bicornis* (Turnbull et al. 1989) and the latter by shifting
prey preferences to springbok, gemsbok and other species (Berry, unpubl. data).
The relationships between rainfall, herbivores and predators in Etosha, where
natural causative factors and anthropogenic influences have combined, are too
complex to make possible long-term predictions at present.

Although anthrax outbreaks in wildebeest and zebra at Etosha can be
predictably linked to the rainy season in both wet and dry phases, it has yet to be
established whether the overall number of herbivores dying from anthrax varies
significantly or is greater during a wet phase than during a dry phase of a "cycle".
Factors which may mask this are the very small proportion of anthrax-infected
carcasses which is actually found and recorded in an area the size of Etosha
(Turnbull 1989), the different levels of susceptibility between herbivore species
(Turnbull et al. 1989) and the continuing changes in total abundance of each
herbivore species.

In the light of the above, Remmert's (1987) first question, namely "What
agents drive the cycles . . . ?" can be answered more or less satisfactorily if, indeed,
cycles are involved in our case histories. In the arid Namib-Naukluft area rainfall
is the primary driving agent, but in the semi-arid Etosha area the system is
additionally driven and made more complex by disease which is largely a
consequence of anthropogenic factors. It may well be that the "mosaic" in Etosha
is changing from a state dominated by long-lived large mammalian herbivores to
one in which conditions are favourable for small organisms whose generation
time is relatively rapid.

The second question posed by Remmert (1987) concerns the size of the
"mosaic stones" and what determines their size. From our two case histories, it
appears that in arid and semi-arid areas the "stones" may cover areas of several
100 km², which is far greater than in tropical forests where mosaics can be
measured in hectares (Remmert 1987). Size determination of natural mosaics in
Namibia may be governed by the primary agents which drive the systems. Since
rainfall is probably the most important but patchy agent in this territory, the size
of mosaics will be accordingly varied and as yet unpredictable.

The effects of rainfall are likely to be especially dramatic in the hyper-arid to arid situation where primary production and herbivore numbers rapidly achieve maximum potential before decreasing to their previous low levels. The implications of this for practical nature conservation are manifold, and special care needs be taken to ensure that Namibian conservation areas are large enough to sustain vital ecological processes.

Acknowledgements. The Weather Office, Windhoek, supplied the isohyetal map and rainfall records. The Desert Ecological Research Unit at Gobabeb made earlier meteorological data available and their Director, Dr. Mary Seely, kindly commented on a draft of this paper. The Secretary of Agriculture and Nature Conservation, Dr. H. Schneider, and the Director of Nature Conservation, Mr. P. Swart, are acknowledged for their support and the Cabinet of the Interim Government of South West Africa/Namibia is thanked for granting H. Berry permission to attend the Mosaic Cycle Hypothesis Symposium in West Germany. We are both very grateful to the Werner Reimers Foundation in Bad Homburg and to Prof. Dr. Hermann Remmert of the Philipps University in Marburg for making it possible for us to travel to the Federal Republic of Germany.

References

Aubreville A (1938) La foret coloniale: les forets de l'Afrique occidentale francaise. Ann Acad Sci Colon Paris 9:1–245

Berry HH (J980) Behavioural and eco-physiological studies on blue wildebeest (*Connochaetes taurinus*) at the Etosha National Park. PhD Thesis, University of Cape Town, South Africa (unpubl)

Berry HH (1981) Abnormal levels of disease and predation as limiting factors for wildebeest in the Etosha National Park. Madoqua 12(4):242–253

Berry HH (1987) Ecological background and management application of contraception in free-living African lions. Proc Congr Contraception in Wildlife, Philadelphia, USA

Berry HH, Louw GN (1982) Nutritional balance between grassland productivity and large herbivore demand in the Etosha National Park. Madoqua 13(2):141–150

Boyer D (1988) The 1988 aerial census of the northern region of the Namib-Naukluft Park. Departmental Report N, 13/4/1/2, Directorate of Nature Conservation, Namibia

Ebedes H (1977) Anthrax epizoötics in Etosha National Park. Madoqua 10(2):99–118

Giess W (1971) A preliminary vegetation map of South West Africa. Dinteria 4:1–114

Hamilton WJ, Buskirk R, Buskirk WH (1977) Intersexual dominance and differential mortality of Gemsbok *Oryx gazella* at Namib Desert waterholes. Madoqua 10(1):5–19

Lancaster J, Lancaster N, Seely MK (1984) Climate of the central Namib Desert. Madoqua 14(1):5–61

Nel PS (1980) Aerial Census October 1980 of the Namib-Naukluft Park (Namib Section). Afrikaans Departmental Report, Directorate of Nature Conservation, Namibia

Pietruszka RD, Seely MK (1985) Predictability of two moisture sources in the Namib Desert. S Afr J Sci 81:682–685

Preston-Whyte RA, Diab RD, Tyson PD (1977) Towards an inversion climatology for southern Africa. Part II: Non-surface inversions in the lower atmosphere. S Afr Geogr J 59:47–59

Remmert H (1987) Sukzessionen im Klimax-System. Verhandlungen der Gesellschaft für Ökologie (Giessen 1986) Band XVI:27–43

Schaller GB (1972) The Serengeti lion. University of Chicago Press, Chicago

Seely MK (1978) Grassland productivity: the desert end of the curve. S Afr J Sci 74:295–297

Seely MK (1987) The Namib — natural history of an ancient desert. John Meinert, Windhoek

Seely MK, Louw GN (1980) First approximation of the effects of rainfall on the ecology and energetics of a Namib Desert dune ecosystem. J Arid Environ 3:25–54

Siegfried WR (1981) The incidence of veld-fire in the Etosha National Park, 1970–1979. Madoqua 12:225–230

Stander PE (1990) Demography of lions in the woodland habitat of Etosha National Park. Madoqua 17 (in press)

Tilson R, Von Blottnitz F, Henschel J (1980) Prey selection by spotted hyaena *Crocuta crocuta* in the Namib Desert. Madoqua 12(1):41–49

Tinley KL (1975) Habitat physiognomy, structure and relationships. The Mammal Research Institute 1966–1975. New Series No (University of Pretoria) 97:69–77

Turnbull PCB (1989) Anthrax in the Etosha National Park. Rössing Magazine (May). Rossing Uranium, Windhoek, pp 1–5

Turnbull PCB, Carman JA, Lindeque PM, Joubert F, Hübschle OJB, Snoeyenbos GH (1989) Further progress in understanding anthrax in the Etosha National Park. Madoqua 16(2):93–104

Tyson PD (1986) Climatic change and variability in southern Africa. Oxford University Press, Cape Town

Weiner J, Gorecki A (1982) Small mammals and their habitats in the arid steppe of central eastern Mongolia. Pol Ecol Stud 8(1/2):7–21

Conclusions

H. REMMERT

1. A natural biotope is occupied by even-aged key organisms (of one or more species); different parts (mosaic stones) of this biotope contain these species in different age groups.

2. Because of the equal age of the key organisms in one mosaic stone, they die at approximately the same time. Very often they are replaced by different — and of course again even-aged — species. Thus, a cycle results of different species or of young — adult — aging — key organisms; these cycles run with the same speed, but not synchronously in a given biotope.

3. As the cycle period is dependent on the normal generation time of the key species, it runs in the same region at approximately the same speed.

4. Very tall (and old) organisms are more sensitive to environmental stress (biotic and abiotic) than young or fully mature organisms. Thus, the highest diversity of a system is found in the aging and collapsing phase of the cycle and is lowest in the growing and early mature phase.

5. Thus, ecosystems cannot be uniform, but always contain mosaic stones with desynchronous cycles. There is no equilibrium in the system, but a constant cyclic succession.

6. The distribution of organisms in a given system is highly clustered.

7. The population pyramid of key organisms in given mosaic stones is extremely different and (in long-lived organisms) is never a "normal" population pyramid.

8. The size of the mosaic stones varies significantly, ranging in forests from a single tree-fall gap (in tropical rain forests or species-rich temperate continental forests) to many square kilometers (in dry tropical forests or taiga-type forests).

9. It is not clear whether the smaller organisms (herbs) in a given system underlie the same type of mosaic cycle as, e.g. in terrestrial systems (trees).

10. There is growing evidence that the cycle with different organisms in one and the same place is of vital importance for modern observations on dieback (*Waldsterben*) in our forests.

Fachbereich Biologie, Universität Lahnberge, Postfach 1929, 3550 Marburg, Germany

H. Remmert (Ed.)
Ecological Studies Vol. 85
© Springer-Verlag Berlin Heidelberg 1991

11. Equilibrium in a supersystem can thus be reached by constant cyclic succession.

12. There are, of course, strong interrelations between mosaic cycles and stochastic patch dynamics.

13. The model predicts high stability of this system against disturbances.

14. The model predicts very low frequency inert oscillations in the supersystem, which are violent if the system is small.

Subject Index

Page numbers in italics refer to figures.